Basic Geological Mapping

Basic Geological Mapping

Editor

Nitin Pradhan

scitus academics

Basic Geological Mapping

Edited by **Nitin Pradhan**

ISBN: 978-1-68117-134-0

Library of Congress Control Number: 2015936555

SCITUS Academics LLC,
www.scitusacademics.com
Box No. 4766, 616 Corporate Way,
Suite 2, Valley Cottage,
NY 10989

Printed in United States of America on Acid Free Paper ∞

Contents

Preface

This book is a basic guide to field techniques used in geological mapping. It is meant to be kept in camp with you and even carried in your rucksack in the field. In addition, because no piece of geological mapping can be considered complete until the geology has been interpreted and explained, chapters are provided on drawing cross-sections, on preparing and presenting 'fair copy' maps, and on presenting geological diagrams from your fieldwork suitable for inclusion in your report. A report explaining the geology is an essential part of any field project, and a brief chapter on the essentials for writing and illustrating it concludes this book. Some emphasis, too, is given to field sketch-mapping because many reports lack those large-scale detailed maps of small areas that can often explain complex aspects of the geology that cannot be shown on the scale of the field map being used, and that are difficult to describe in words. Attention is also given to field notebooks, which are, in many cases, deplorable.

Editor

Chapter 1

Seismic Microzonation of Breginjski Kot (NW Slovenia) Based on Detailed Engineering Geological Mapping

Jure Kokošin[1] and Andrej Gosar[2, 3]

[1]Sweco Norge AS, Dronningensgate 52/54, 8509 Narvik, Norway

[2]Faculty of Natural Sciences and Engineering, University of Ljubljana, Ašker eva 12, SI-1000 Ljubljana, Slovenia

[3]Slovenian Environment Agency, Seismology and Geology Office, Dunajska 47, SI-1000 Ljubljana, Slovenia

ABSTRACT

Breginjski kot is among the most endangered seismic zones in Slovenia with the seismic hazard assessed to intensity IX MSK and the design ground acceleration of 0.250 g, both for 500-year return period. The most destructive was the 1976 Friuli Mw = 6.4 earthquake which had maximum intensity VIII-IX. Since the previous microzonation of the area was based solely on the basic geological map and did not include supplementary field research, we have performed a new soil classification of the area. First, a detailed engineering geological mapping in scale 1:5.000 was conducted. Mapped units were described in detail and some of them interpreted anew. Stiff sites are composed of hard to medium-hard rocks which were subjected to erosion mainly evoked by glacial and postglacial age. At that time a prominent topography was formed and different types of sediments were deposited in valleys by mass flows. A distinction between sediments and weathered rocks, their exact position, and thickness are of significant importance for microzonation. On the basis of geological mapping, a soil classification was carried out according to the Medvedev method (intensity increments) and the Eurocode 8 standard (soil factors) and two microzonation maps were prepared. The bulk of the studied area is covered by soft sediments and nine out of ten settlements are situated on them. The microzonation clearly points out the dependence of damage distribution in the case of 1976 Friuli earthquake to local site effects.

INTRODUCTION

The Breginjski kot (NW Slovenia, Figure 1) is located close to a seismically very active area of Friuli in NE Italy, but also the Julian Alps in Slovenia have recently experienced an increase of seismic activity. In this area, in fact, the seismic hazard in Slovenia is the highest. Most of numerous settlements in the area are built on soft sediments which can significantly enhance the ground shaking. The whole area was highly damaged during the Friuli 1976 Mw = 6.4 earthquake and its aftershock sequence. For a realistic assessment of site effects and earthquake hazard, a thorough seismic microzonation is therefore essential. On the other hand the existing microzonation of Breginjski kot [1] was based solely on the basic geological map of Slovenia in scale 1: 100.000, without supplementary field investigations or subsurface information, thus making it fairly inaccurate. Therefore we have decided to perform a new microzonation study based on a detailed engineering geological mapping in scale 1: 5.000. Such a map would enable soil classification according to two different approaches without using expensive geotechnical drilling or geophysical investigations. The seismic microzonation according to Eurocode 8 was based on a new European seismic standard [2] which has been applied in Slovenia from 2008. This standard defines the seismic hazard by the peak ground acceleration, whereas the site effects are expressed by the soil factor. The product of both values is used in the design of earthquake resistant buildings. The seismic microzonation according to the Medvedev method [3] denotes seismic hazard by increments of seismic intensity. Although the seismic intensity in Slovenia is not used for building design anymore, it is still important for seismological analyses and for the civil protection purposes.

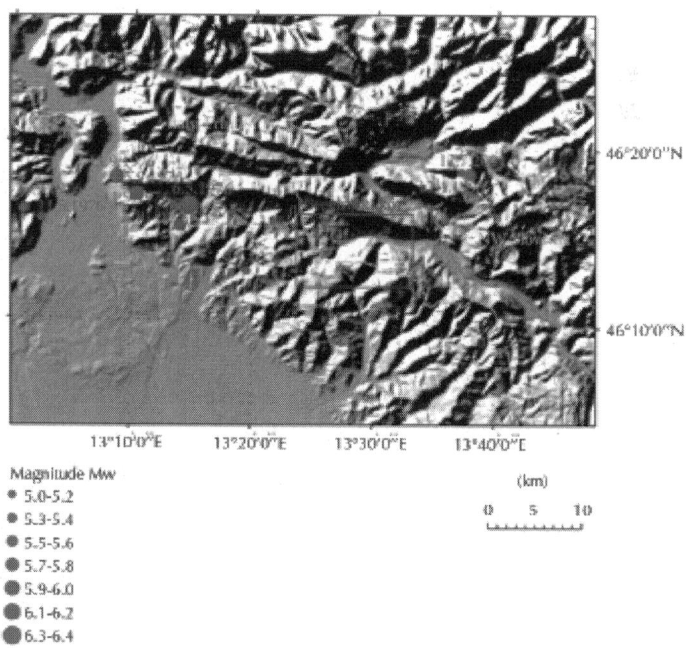

Figure 1: Seismicity (Mw ≥ 5.0) in Friuli and NW Slovenia with location of Breginjski kot study area.

SEISMOLOGICAL CHARACTERISTICS

The Breginjski kot is located in one of the three areas with the highest seismic hazard in Slovenia. This is mainly due to the proximity of the seismically very active area of Friuli located 10–30 km to the west in NE Italy. In this area the Mw = 6.4 Friuli earthquake occurred in May 1976. The seismic sequence that followed (Table 1, Figure 1) consisted of six events of Mw between 5.0 and 6.0 during the same year and one event of Mw = 5.3 that occurred in September 1977 [4, 5]. The intensity of the main shock in Breginj was VIII EMS-98, but the cumulative intensity which includes also damage from aftershocks reached IX EMS-98. The highest intensity in Slovenia from the main shock was VIII-IX in Robidiš e [6]. A complete review of 1976 earthquake intensities in Breginjski kot settlements is summarized in Table 2.

Table 1: Data for earthquakes with Mw ≥ 5.0 in the vicinity of Breginski kot (after [4, 7,8])

Date	**Time (UTC)**	**Lat (N)**	**Lon (°N)**	**Depth (km)**	**Mw**	**Region**
06/05/1976	20:00	46.275	13.246	6.5	6.4	Friuli
09/05/1976	00:53	46.214	13.326	8.5	5.1	Friuli
11/05/1976	22:44	46.260	13.041	6.0	5.0	Friuli
11/09/1976	16:31	46.275	13.224	6.5	5.2	Friuli
11/09/1976	16:35	46.256	13.233	4.3	5.6	Friuli
15/09/1976	03:15	46.284	13.173	5.0	5.9	Friuli
15/09/1976	09:21	46.300	13.145	7.0	6.0	Friuli
16/09/1977	23:48	46.268	13.016	8.0	5.3	Friuli
12/04/1998	10:55	46.309	13.632	7.6	5.6	Krn Mts.
12/07/2004	13:04	46.310	13.620	11.0	5.2	Krn Mts.

Table 2: Intensities of 1976 earthquake in different settlements of Breginjski kot

Settlement	Intensity EMS-98
Borjana	VI
Kred	VI
Robi	VI
Potoki	VI+ (MSK-78)
Logje	VI-VII
Staro Selo	VII
Sedlo	VII-VIII
Breginj	VIII
Podbela	VIII
Robidiš e	VIII-IX

For the NW Slovenia relatively weak rates of seismicity were observed before April 1998 when a Mw = 5.6 earthquake occurred in Krn Mountains, only 7 km NW from Kobarid [9, 10]. It was followed by a Mw = 5.2 event in July 2004 in the same epicentral area (Figure 1). Both earthquakes occurred on the NW-SE trending near-vertical dextral strike-slip Ravne fault [11] in the depth range 7.6–11 km. The intensity of the 1998 earthquake in Kobarid (located at eastern margin of the Breginjski kot) was VI-VII EMS-98 [7] and of the 2004 event VI EMS-98 [8].

NW Slovenia and Friuli region are located at the kinematic transition between E-W striking thrust faults of the Alpine system (Friuli earthquakes) and NW-SE striking strike-slip faults of the Dinarides system (Krn Mountains earthquakes) (Figure 1). The strongest earthquake ever recorded in the Alps-Dinarides junction area was the 1511 western Slovenia earthquake (Mw = 6.8); the exact location and mechanism of this event are still debated [12, 13] due to early occurrence and thus very little or no historical documents.

According to the old seismic hazard map of Slovenia for a 500-year return period showing expected intensities on the MSK scale [14], Breginjski kot is estimated to have IX MSK intensity. According to the new seismic hazard map for a 475-year return period [15], a design ground acceleration value for a rock site in Breginjski kot is 0.250 g. The important methodological difference between both maps is that intensities are assessed for a "medium" soil type, whereas design ground acceleration is assessed for a "rock" site according to Eurocode 8.

Accelerographs were installed in Kobarid after the main Friuli shock in 1976 and after the Krn Mountains earthquake in 1998. The strongest ground motion in the Friuli seismic sequence was recorded for the September 15, 1976 aftershock (Mw = 6.0, distance = 37 km) as 0.138 g peak ground acceleration [16]. For the July 12, 2004 Krn Mountain earthquake (Mw = 5.2, distance = 7 km) a peak ground acceleration of 0.152 g was recorded [17].

The previously existing microzonation of Breginjski kot [1] was based solely on the basic geological map in scale 1 : 100.000, without supplementary field research or subsurface geological and geophysical information, thus making it fairly inaccurate. It was prepared to be used together with the old seismic hazard map showing intensities on the MSK scale. This microzonation showed that the maximum expected intensity due to the effects of soft sediments can be increased by up to one intensity degree.

ENGINEERING GEOLOGICAL MAPPING

A detailed engineering geological map of Breginjski kot in scale 1 : 5.000 was prepared by outcrop observing and geological boundaries tracking method. Since seismic microzonation was the final goal, the main focus of the field survey was to characterise local geological conditions in vicinity of settlements and other substantial infrastructure like roads, power lines, water collectors, and so forth. Consequently some less relevant lithological boundaries were summarised from the manuscript geological map in scale 1 : 25.000, which was the basis for published basic geological map in scale 1 : 100.000 [18]. Based on the mapping accomplished, rocks and sediments were classified by their characteristics determined in the field. Properties such as structure, dip, strength and weathered material thickness were defined for rocks, whereas granulation, roundness, and thickness were described for unconsolidated sediments. Mapped lithological units were compared with the Bovec basin, located 15 km to the NE in the So a valley for which intensity increment microzonation was performed after the 1998 earthquake [19]. Lithologies described below (Figures 2 and 3) are arranged by age order and are presented on a detailed engineering geological map (Figure 4) of Breginjski kot. Their names and age classification mainly coincide with lithologies described in the explanatory text of the basic geological map of Slovenia [20].

(a)

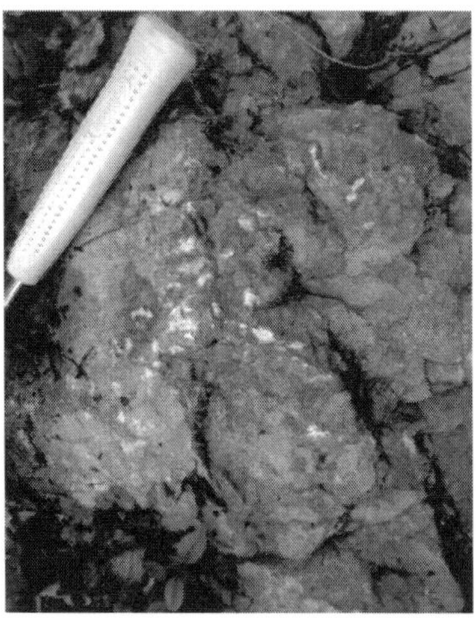

(b)

(c)

(d)

(e)

(f)

Figure 2: Typical lithologies according to the engineering-geological mapping: (a) Megalodontidae shell in Dachstein limestone in Robi ,

(b) limestone breccia interbedded with shalestone near Robidiš e, (c) folds in flyschoid formation near Borjana, (d) heterogeneous composition of diamicton in Sedlo, (e) lacustrine chalk SW of Podbela, and (f) proluvium in Podbela.

(a)

(b)

(c)

(d)

Figure 3: (a) Karst sinkhole in coral limestone represents a depositional place for significant amount of weathered material; Robidiš e in the background. (b) Limestone breccia in Nadiža valley south of Logje is typically bedded. (c) Diamicton near Breginj, in the background Mt. Stol built of Dachstein limestone. (d) Alluvial fan sediments form gentle slopes above Kred polje, which is composed of fine grained alluvial sediments which promote to alluvial sediments near Nadiža River channel.

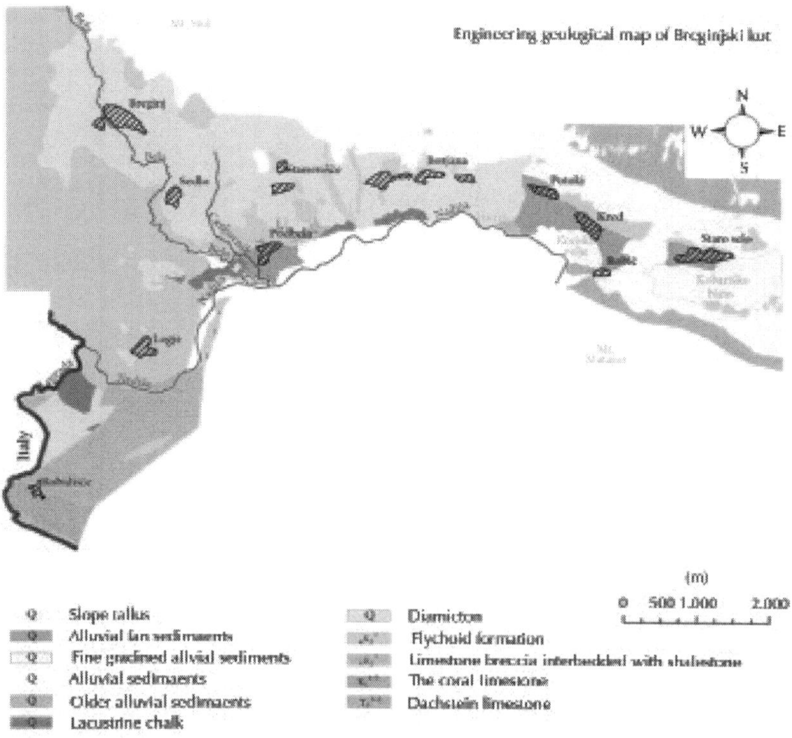

Figure 4: Engineering geological map of Breginjski kot.

Dachstein limestone (Figure 2(a)) from Late Triassic is the oldest formation in the studied area. It forms Mt. Stol, Mt. Matajur, and Der Hill west of Robi . Rock has a light grey colour and a crystalline texture. Dachstein limestone is characterized by Lofer development which consists of three parts. Cyclothem of B member contains stromatolites whereas cyclothem of C member contains well-preserved Megalodontidae shells. Cyclothem of A member consists of basal breccias but was not observed in the field. Weathered material thickness was estimated to be less than 0.5 m.

The Coral limestone of Early Cretaceous is situated in a small part of the examined area near Robidiš e. Coral limestone is fine grained, massive and grey in colour. Outcrops are rare because they are overlain by significant thickness of weathered material. Hence it was distinguished from limestone breccias by numerous karst sinkholes which determine its morphology (Figure 3(a)). Moreover the weathered material thickness can be altered promptly as a consequence of a karst relief. On the slopes less than 1 m of weathered material is expected, whereas more than 5 m of weathered material is estimated for the flatland.

Limestone breccia interbedded with shalestone (Figure 2(b)) from Upper Cretaceous represents mainly the area from Napoleon Bridge west of Podbela to Robidiš e (Figure 3(b)). Limestone breccias are of light grey or grey colour. Breccia is clast supported and composed of angular to subangular clasts which range from 2 mm to 1 m in size. Breccia formation is mostly bedded, rarely massive. If bedded, its thickness ranges from 10 cm to several meters in size, moreover brownish green shalestones are often interbedded. Formation generally dips to the north. On the slopes less than 1 m of weathered material is expected, whereas from 2 m to 3 m of weathered material is estimated for the flatland.

Flyschoid formation (Figure 2(c)) is the next formation of Upper Cretaceous and is situated mainly west of Breginj-Sedlo-Podbela line, while plenty outcrops are exposed in erosion windows of Mt. Stol and Mt. Matajur, especially in ravines with high-energy creeks. The Flyschoid formation is divided into series of rocks: shalestone, sandstone, and breccia, which are stratified and in sequence one after the other. Shalestone is dark grey and appears with thickness of beds ranging from 1 cm to 5 cm. Sandstone, which mostly consists of carbonate minerals, is grey with thickness of beds ranging from 5 cm to few ten centimetres. Breccia is grey to dark grey and consists of carbonate clasts ranging from 0.5 cm up to few meters. Coarse grained breccia is clast supported. Thickness of breccia beds ranges from 1 m to few ten meters. Flyschoid formation dips generally towards the north, moreover thin bedded series in particular are often found folded. Thickness of weathered material varies from 0.5 m to 1 m on the slopes and from 2 m to 3 m on the flatland.

Diverse Quaternary sediments in Breginjski kot are the result of postglacial era when water erosion and other transformational processes formed specific relief. Sediment deposits are mostly abundant on the slopes of Mt. Stol where most settlements are located.

Diamicton (Figure 2(d)) is an unlithified sediment which origins from glacial till mixed with other Quaternary sediments and transported by mass flow. It is deposited on the slope of Mt. Stol (Figure 3(c)), confined by Potoki to the east, a large area of flyschoid formation to the west and Nadiža River to the south. In addition some erosion patches can be found between Sedlo and Logje. According to the estimation on the field, diamicton is composed of more or less subrounded grains which are very poorly sorted. 70% of grains represent fine gravel, sand, and silt fractions, which ratio can vary significantly on studied locations. The remaining 30% is mainly gravel which ranges from 1 cm to 5 cm in size and boulders which ranges from 20 cm to several meters in size. Fragments consist of carbonate in general. Besides, there are many clastic fragments and diamictite ones, which are rather scarce. Glacial striations on fragments were not observed. Diamicton thickness varies significantly, for instance: along the Nadiža River only a few meters can be reached, in vicinity of flyschoid formation erosion windows it can vary between 5 m and 10 m, whereas away from ravines it can reach more than 20 m. The thickness depends on paleorelief and selective sedimentation processes, therefore a large amount of material has been deposited there. For instance in Cerkovnik creek flyschoid formation is not exposed, despite more than 20 m deep eroded ravine. According to ravine morphology, the biggest thickness is expected in the vicinity of Breginj.

Diamictit has the same origin as diamicton but it is lithified. It is not presented on the engineering-geological map, because it is scattered around the Diamicton area in small patches of about 10 m^2 in size. It consists of around 70% of gravel and boulder sized clasts which are matrix supported. The latter is composed of silt, sand and fine gravel sized grains. Subrounded clasts are poorly sorted, those in the range from 0.5 cm to 5 cm prevail. Glacial striations on clasts are rarely preserved.

Lacustrine chalk (Figure 2(e)) is deposited in patches along broader zone of Nadiža River and in the largest portion besides Legrada creek. This sediment is characterised by laminated silt, grey in colour with a blue tinge which has moisture and can be slightly molded. According to Casagrande's Soil Classification System (ASTM) [21] it was defined as ML, inorganic silt and very fine sands with low plasticity. Besides laminated silt there can be less than 5% of rounded pebbles found, mainly ranging from 1 cm to 5 cm and rarely to 10 cm. Lacustrine chalk is covered by diamicton; main evidence can be found near to Hurja farm.

Older alluvial sediments can generally be found between Kred and Podbela up to 320 m above sea level as erosion patches around 40 m^2 in size and are consequently not thicker than 5 m. They include conglomerate, but sands and gravel are more rare. First is clast supported conglomerate which is good sorted. The clasts, which range from 0.2 to 5 cm in size, are well rounded and consist mainly of carbonate and flyschoid fragments.

Alluvial sediments are located along former and actual steam channels of Nadiža River and Bela creek and in alluvial terraces of Kred polje (Figure 3(d)) and Podbela area. The volume ratio of sand and pebbles is 60 : 40. Fragments, which consist of carbonate and flyschoid formations, are well-rounded and moderately sorted. Thickness of alluvium deposits is estimated to be more than 10 m.

Fine grained alluvial sediments are deposited in Kred polje (Figure 3(d)) and Kobariško blato which represent a sedimentary basin between Mt. Stol and Mt. Matajur. The flat relief induces rare outcrops, whereas some exposures were seen in old melioration trenches. Some 4 m pits, which were excavated in the frame of geological investigations for planned wastewater treatment plant, were also examined. A particular sediment found in trenches and pits is characterised by silt of bluish grey colour. According to ASTM it is defined as ML, inorganic silt with low plasticity. On the foothills of Mt. Stol the formation is overlain by alluvial fan sediments.

Alluvial fan sediments (proluvium) (Figure 2(f)) are deposited by torrential creeks on the foothills of Mt. Stol from west of Staro Selo to Potoki (Figure 3(d)). In Podbela alluvial fan sediments are found as well. The formation consists of flyschoid and carbonate fragments, whereas the ratio depends on geological setting of hinterlands. Proluvium is poorly sorted since it consists of fragments generally ranging up to 10 cm, of which up to 40% is fine sand and silt. Boulders ranging up to 60 cm can also be observed. In addition, fragments are medium rounded. Thickness of formation is estimated on less than 20 m.

Slope talus (gravel) is deposited below steep slopes of Mt. Stol and Mt. Matajur. The lower boundary of the slope talus on the slope of Mt. Stol is increasing in elevation from east to west, whereas on the slope of Mt. Matajur it is decreasing in the same direction. Gravel consists of limestone and dolomite fragments of Late Triassic. They were distinguished by the reaction with dilute HCl. Besides gravel sized fragments, also significant amount of cobbles and some boulders can be found. Additionally patches of lithified slope tallus (gravel breccia), which have the same origin as unlithified, can be observed, but they are not presented in the map due to their negligible size and dispersion.

The described lithologies are presented on the detailed engineering geological map (Figure 4), which is the basis for the seismic microzonation of the studied area.

INTENSITY INCREMENTS SEISMIC MICROZONATION

According to the old seismic hazard map of Slovenia for a 500-year return period [14] Breginjski kot is estimated to have IX MSK intensity. It is well-known that variations in local site characteristics have influence on observed earthquake intensity. Medvedev [3] proposed a classification that improves estimation on intensity considering engineering geological conditions. According to this classification, the soil is divided into three categories. The 2nd category corresponds to the medium stiff ground, which agrees with the given intensity on the seismological map [14], whereas for the 1st category an intensity decrease, and for the 3rd category an intensity increase is expected. According to this scheme Breginjski kot is subdivided into three categories (Table 3) and presented in the map (Figure 5). In addition, engineering geological conditions, topography, and soil saturation are given descriptively in the text below, according to Medvedev [3] and Mayer-Rosa and Jimenez [22].

Table 3: Soil categories for seismic microzonation according to Medvedev method (after [3])

Soil category	Soil type	Intensity increment according to seismic area		
		VII	VIII	IX
I	hard rocks	VI	VII	VIII
II	medium-hard rocks, lithified Quaternary sediments, alluvial sediments	VII	VIII	IX
III	unlithified Quaternary sediments	VIII	IX	X

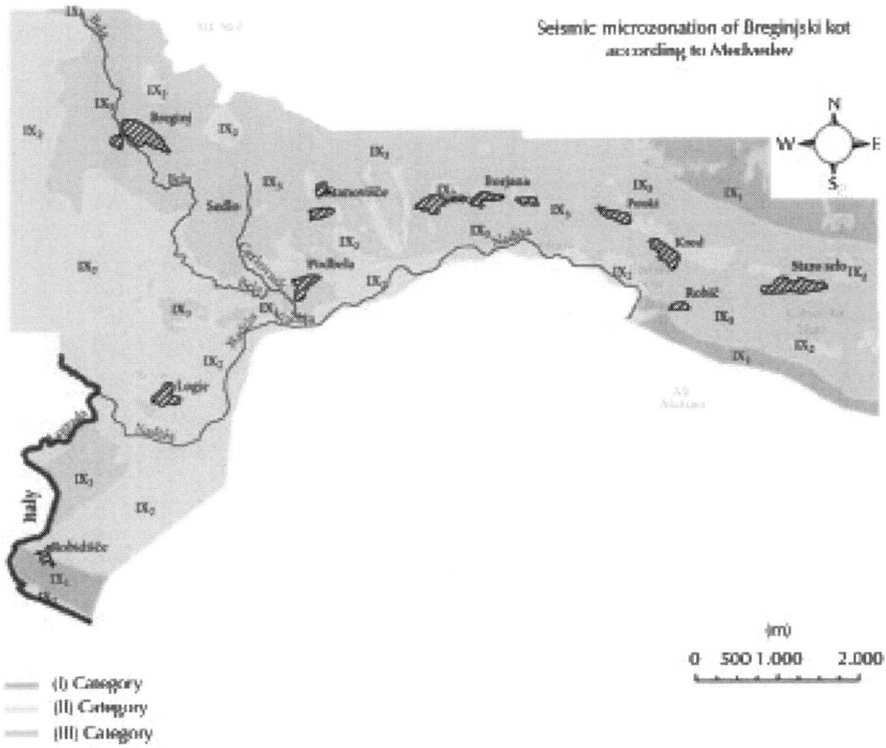

Figure 5: Intensity increments seismic microzonation of Breginjski kot according to the Medvedev method.

Category I represented by Dachstein limestone and coral limestone which are in general seismically insensitive. But, due to limestone karstification some intensity increments may be expected also for these rocks. In addition, terrain subsidence is expected because of cave collapses. Severe earthquakes may also trigger rockfalls and rock slides which can endanger settlements or road facilities.

Category II is characterised by flyschoid formations, limestone breccia interbedded with shalestone, older alluvial sediments, and alluvial sediments. Besides them, diamictit and slope tallus (gravel breccia) are also within this category, but are not presented on the map because of negligible size and dispersion. Intensity increments may be significant due to the high- degree saturation of soil along Nadiža River or Bela creek and limestone breccia karstification. Severe earthquakes may also trigger rockfalls and rock slides which can endanger settlements or road facilities. Areas with increased thickness of weathered material are prone to landside occurrences. Many caves, abysses, and sinkholes are observed in the Robidiš e plateau (Figure 3(a)), where the thickness of weathered material in general exceeds 5 m and can even reach 20 m in sinkholes. A local increment of intensity can be therefore expected.

Category III is represented by lacustrine chalk, fine grained alluvial sediments, alluvial fan sediments (proluvium), diamicton, and slope tallus (gravel). Intensity increments might be larger at contacts between different sediments, especially in Podbela, Krejsko polje, and Kobariško blato which are located on the foothills of Mt. Stol. Moreover, prominent topography changes like in Sedlo, which is situated in crest vicinity, can cause an additional intensity increment.

Since significant diversities in geological conditions and topography are observed in the studied area, which might cause an increase of intensity, five main reasons for site effects in different settlements of Breginjski kot are given in Table 4. Soft sediments should be considered as a complex set of depositional events of distinctive origin, hence different sediment layers exist, divided by a certain number of sediment contacts. Generally, the larger the number of contacts is, the larger the intensity increment is. In addition, prominent topography with steep slopes, sharp relief changes and abundant karst features are also reasons for an intensity increase.

Table 4: Main reasons for significant site effects in settlements of Breginjski kot

Settlement	Main reasons for site effects				
	Prevailing lithology	**No. of sediment contacts**	**Thickness to the basement rock**	**Topography**	**Karst features**
Borjana	Diamicton	1	5–10 or more	Steep-moderate slopes	no
Kred	Proluvium	2	<10 m	Flat	no
Robi	Gravel and limestone	2	<5 m	Flat	no
Potoki	Proluvium and flyschoid f.	1	0–10 m or more	moderate slopes	no
Logje	Flyschoid f.	0*	<2 m	slight slopes-flat	no
Staro Selo	Proluvium	3	>20 m	slight slopes-flat	no
Sedlo	Diamicton	1	10–20 m or more	slight slopes, sharp relief change	no
Breginj	Diamicton	1	>20 m	slight slopes-flat	no
Podbela	Proluvium	2	>20 m	flat	no
Robidiš e	Limestone and limestone breccia	0*	>5 m	flat	yes

*The thickness of weathered material is estimated in text.

Since 2008 seismic intensity is not used in Slovenia for building design anymore, but it is still important for seismological analysis and for civil protection. Broader applicability of the method is also in comparison with similar studies. The new seismic microzonation according to Medvedev method improves the older seismic microzonation [1] in many aspects: (a) mapping was performed on a larger scale, (b) lithologies descriptions are more precise and new interpretations are made and (c) map is digitalized. Moreover, the Bovec basin seismic microzonation [19] can also be compared to

Breginjski kot due to geographic proximity and similar geological structures, though many different geotechnical and geophysical data were also acquired in the Bovec basin, as well as a study based on the microtremor method [23].

EUROCODE 8 SEISMIC MICROZONATION

The new seismic hazard map of Slovenia defines the seismic hazard with the peak ground acceleration. Furthermore, local ground conditions account for the additional influence on the seismic ground motion. In Europe ground types with different soil factors are usually determined according to the seismic standard Eurocode 8 [2] and National Annex [24]. According to the seismic hazard map for a 475-year return period [15], Breginjski kot is characterized in its larger western part by a design ground acceleration value for a rock site of 0.250 g and in the smaller easternmost part by 0.225 g. Ground types are classified on the basis of their geomechanical properties and weathered material thickness. Thus four different ground types, namely A, C, D, and E were assessed for the studied area (Figure 6).

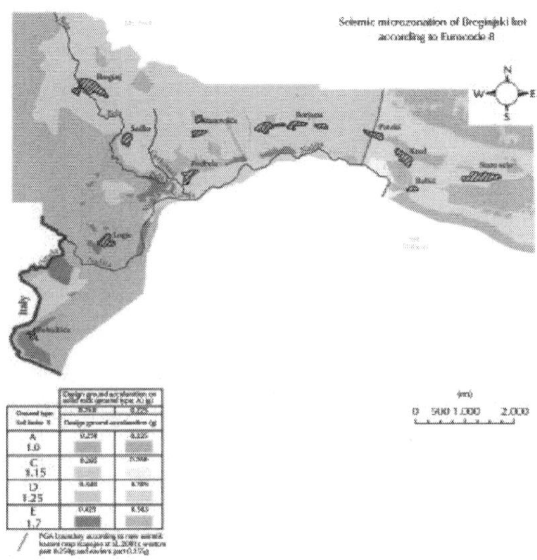

Figure 6: Eurocode 8 seismic microzonation of Breginjski kot.

Class A represents the least vulnerable ground type. It is characterised by limestone, limestone breccia interbedded with shalestone and flyschoid formation, and less than five meters of weathered material. According to the map (Figure 6) the concentration of class A is the largest in the westernmost part, though small patches are present throughout the area. Logje is the only settlement which is located entirely on class A ground. Only small parts of Robič, Potoki, and Borjana are lying on class A ground.

Class C is characterized by alluvial sediments, several tens of meters thick. The area which corresponds to the class C is situated along Nadiža River or Bela creek where no settlement is located.

Class D is characterized by diamicton, alluvial fan sediments, and slope tallus (gravel). This is the largest area in the map, which is mainly located on the slopes of Mt. Stol and Mt. Matajur. In addition it can be found in two patches between Sedlo and Robidišče. Due to its wide extent, most settlements belong to this class, for example, Staro Selo, Kred, Robič, Potoki, Borjana, Podbela, Sedlo, and Breginj.

Class E has the highest soil factor according to Eurocode 8. It comprises soil type C or D with a thickness varying from 5 m to 20 m underlain by a rock of type A. In addition, fine grained alluvial and lacustrine sediments belong to the latter type as a result of their poor geomechanical properties and heterogeneity. The area of class E includes Kobariško blato, Krejsko polje, Robidiš e plateau, and some smaller patches along Nadiža valley. Robidiš e is the only settlement which corresponds to the class E.

In addition to the upper classification some additional risks persist for each class, such as terrain subsidence in areas covered by limestone due to karst caves collapse and possibility for rockfalls. Final design acceleration values for each ground type are computed in accordance to the following expression:

$$\mathrm{DGA_F} = \mathrm{DGA_A} \times S \times T_{\mathrm{NCR}}, \tag{1}$$

where final design ground acceleration (DGA_F) corresponds to design ground acceleration on type A ground spectral acceleration multiplied by the soil factor (S) and the factor for the reference return period (T_{NCR}). Topographic factor is neglected in this study but it should be taken into account for important structures according to Eurocode 8 [25]. Final design ground acceleration values are shown in Table 5.

According to the National Annex [24] the soil factor for ground type E is 1.7, whereas in the original Eurocode 8 standard [2] it is 1.4; for all other ground types the soil factors are the same.

Table 5: Final design ground acceleration values for each soil type in Breginjski kot

Soil type		T_{NCR}	DGA_A [g]	
			0.250	**0.225**
			DGA_E [g]	
A	1.0	475 years	0.250	0.225
C	1.15		0.285	0.260
D	1.35	1.00	0.340	0.305
E	1.70		0.425	0.383
A	1.0	1000	0.315	0.280
C	1.15	years	0.360	0.325
D	1.35	1.25	0.420	0.380
E	1.70		0.531	0.478

DISCUSSION OF RESULTS

According to the engineering geological map of Breginjski kot, twelve different lithologies are presented and described in detail. Several of them are interpreted anew compared to previous studies (e.g., [18]).

Since typical glacial sediments are missing due to mass flows, alternative term diamicton for pseudoglacial sediments is suggested. Diamictit is also presented but only locally and in small patches. A recent study has shown that similar types of soft sediments or rocks are also identified in nearby Bovec basin [26]. Furthermore, gravel breccia was also observed in small patches similar to diamictit. Neither of them are presented in the engineering geological map but are only presented descriptively, because of their small size and sparse occurrence.

Fine grained alluvial sediments in Krejsko polje and Kobariško blato were deposited after Nadiža River had changed the current direction from the relatively wide Kobariško blato valley into the recent Nadiža valley, which is very narrow, with steep and rocky slopes [20]. Hence relatively large area was covered by swamplands and oxbow lakes where fine grained material was deposited and accumulated throughout Quaternary, although intense melioration in modern ages results in controlled river channel and thus in decreased sedimentation.

Because of detailed mapping, lithologies are defined more accurately and restricted in comparison with previous studies [20]. Therefore, their position and geometry can be better understood, thus contributing to the adequate estimation of the number of sediment contacts, their thicknesses, and abundance of karst phenomena, which may result in additional site effects (Table 4). Moreover, thicknesses of weathered material were estimated for each rock formation, because this also has influence on seismic effects.

A significantly large thickness of weathered material is located in the Robidišče plateau due to extreme weathering of shalestone which outcrops at the edge of the plateau. The Robidiš e plateau is characterized by limestone breccia and coral limestone, which includes karst features like sinkholes where the thickness of weathered material generally exceeds 5 m. Moreover, the thickness of weathered material varies significantly, since in some areas bedrock outcrops are observed. Karst phenomena like caves and abysses can also contribute to an increase of seismic ground motion. According to the geological mapping, limestone breccia interbedded with shalestone is less subjected to karst feature occurrences or such features are scarce. In the settlements of Breginjski kot small thickness of weathered material was observed especially in Logje, where outcrops of rock formation are abundant. On the other hand, the outcrops in Borjana, Potoki, and Robi were limited, thus soft sediments are more abundant. Their cumulative thickness of less than 5 m is expected. However, it should be noted that geological characteristics of sediments and their thickness are estimated from the engineering geological mapping, whereas for a more detailed analysis systematic geotechnical drilling and geophysical investigations should be performed.

The damage to buildings in settlements of Breginjski due to the 1976 Friuli earthquake was unevenly distributed, although their distance from the epicentre does not vary significantly. For comparison, the soil classification is determined for each settlement of Breginjski kot according to both seismic microzonations. A comparison between intensities of the 1976 earthquake and soil classifications is given in Table 6.

Table 6: Intensities of 1976 earthquake in different settlements of Breginjski kot and soil classification after Medvedev and Eurocode 8 from both seismic microzonation maps

Settlement	Intensity EMS-98	Soil classification	
		After Medvedev	After Eurocode 8
Borjana	VI	III, II	D, A
Kred	VI	III	D
Robi	VI	III, I	D, A
Potoki	VI+ (MSK-78)	III, II	D, A
Logje	VI-VII	II	A
Staro Selo	VII	III	D
Sedlo	VII-VIII	III	D
Breginj	VIII	III	D
Podbela	VIII	III	D
Robidiš e	VIII-IX	II	E

Soil classification according to Medvedev method does not show a clear correlation of ground type to observed intensities. The main reason for this is that Medvedev soil classification does not take into account the thickness of soft sediments or weathered material. In addition it does not consider sufficiently engineering geological conditions which are annotated descriptively only. On the contrary, Eurocode 8 soil classification clearly shows a correlation between the ground type and observed intensities. This standard classifies ground types more precisely, also since it considers the thickness of soft sediments and weathered material. A better evaluation of local soil effect is therefore achieved. Based on the 1976 earthquake intensities, the settlements of Breginjski kot can be divided into two categories. The first group consists of settlements which show relatively low site effects like Borjana, Kred, Robič, Potoki, and Logje. They are located either on ground type A, or on the weathered material less than 5 m thick. The indicators of small weathered material thickness are frequent erosion windows. The second group consists of settlements which show relatively high site effects like Staro Selo, Sedlo, Breginj, Podbela, and Robidiš e. They are located on soft sediments with considerable thickness. Additional causes of significant site effects in each particular settlement are described in Table 4.

CONCLUSIONS

This study represents a contribution to the evaluation of the seismic hazard in Breginjski kot, which is among the most endangered seismic areas in Slovenia. Two new seismic microzonations were prepared, because the previous seismic microzonation was based solely on the basic geological map and did not include supplementary field research, thus making it fairly inaccurate. In this study detailed engineering geological mapping of Breginjski kot in scale 1: 5.000 was conducted. The results of mapping showed a diverse composition of sediments and rocks. Mapped units were described in detail and some of them are interpreted anew. Descriptions of mapped units also include the estimated thickness of soft sediments and weathered material. The results can be enhanced and validated by drilling and geophysical investigations, which is recommended especially in the settlements that were badly damaged during the 1976 Friuli earthquake. In addition, a microtremor method study would be appropriate to apply in the flatlands of Krejsko polje and Kobariško blato to determine soil resonance frequencies.

On the basis of engineering geological mapping, seismic microzonation was prepared according to the Medvedev method and the Eurocode 8 standard. According to the first classification the soil in Breginjski kot was divided into three categories, and according to the latter classified into categories A, C, D, and E. Two maps of seismic microzonation, based on both soil classifications, were made in the GIS platform and are therefore ready to use in urban planning. Seismic microzonation according to Eurocode 8 is more appropriate for the application, because it is in accordance with the modern seismic hazard standards and because it better correlates thickness of soft sediments or weathered material with site amplification. Moreover, ground types are classified more precisely, and thus the correlation with the geological map is better. The new seismic microzonation reveals the correlation of site effects and observed intensities. Sites located on softer ground experienced a higher degree of damage during the Friuli 1976 earthquake. Although the microzonation based on seismic intensities is not used in earthquake engineering anymore, it is still important for civil protection purposes and in seismological analysis. The new map improves the old microzonation and enables a comparison of geologically similar areas of Breginjski kot and the Bovec Basin.

ACKNOWLEDGMENTS

The study was realized with the support of the research program P1-0011 financed by Slovenian Research Agency.

REFERENCES

1. R. Vidrih, M. Godec, and J. Lapajne, Earthquake Risk in Slovenia, Seismological Survey of Slovenia, Ljubljana, Slovenia, 1991.
2. CEN, Eurocode 8—Design of Structures for Earthquake Resistance, Part 1: General Rules, Seismic Actions and Rules for Buildings, European Standard, EN, 1998-1: 2004 (E), Stage 64, European Committee for Standardization, Brussels, Belgium, 2004.
3. S. V. Medvedev, Engineering Seismology, Građevinske knjige, Belgrade, Serbia, 1965.
4. B. Perniola, G. Bressan, and S. Pondrelli, "Changes in failure stress and stress transfer during the 1976–77 Friuli earthquake sequence," Geophysical Journal International, vol. 156, no. 2, pp. 297–306, 2004.
5. G. B. Carulli and D. Slejko, "The 1976 Friuli (NE Italy) earthquake," Giornale di Geologia Applicata, vol. 1, pp. 147–156, 2005.
6. V. Ribarič, "Earthquakes in Friuli in 1976 and the short history of seismicity at the margin of Eastern Alps," Potresni Zbornik, pp. 17–80, 1980.
7. P. Zupančič, I. Cecić, A. Gosar, L. Placer, M. Poljak, and M. Živčić, "The earthquake of 12 April 1998 in the Krn Mountains (Upper Soča valley, Slovenia) and its seismotectonic characteristics," Geologija, vol. 44, pp. 169–192, 2001.

8. I. Cecić, M. Živčić, T. Jesenko, and J. Kolar, "Earthquakes in Slovenia in 2004," Potresi V Letu 2004, pp. 16–40, 2006.
9. J. Bajc, A. Aoudia, A. Saraò, and P. Suhadolc, "The 1998 Bovec-Krn mountain (Slovenia) earthquake sequence," Geophysical Research Letters, vol. 28, no. 9, pp. 1839–1842, 2001.
10. A. Gosar, "Site effects and soil-structure resonance study in the Kobarid basin (NW Slovenia) using microtremors," Natural Hazards and Earth System Sciences, vol. 10, no. 4, pp. 761–772, 2010.
11. V. Kastelic, M. Vrabec, D. Cunningham, and A. Gosar, "Neo-Alpine structural evolution and present-day tectonic activity of the eastern Southern Alps: the case of the Ravne Fault, NW Slovenia," Journal of Structural Geology, vol. 30, no. 8, pp. 963–975, 2008.
12. F. Fitzko, P. Suhadolc, A. Aoudia, and G. F. Panza, "Constraints on the location and mechanism of the 1511 Western-Slovenia earthquake from active tectonics and modeling of macroseismic data,"Tectonophysics, vol. 404, no. 1-2, pp. 77–90, 2005.
13. R. Camassi, C. H. Caracciolo, V. Castelli, and D. Slejko, "The 1511 Eastern Alps earthquakes: a critical update and comparison of existing macroseismic datasets," Journal of Seismology, vol. 15, no. 2, pp. 191–213, 2011.
14. V. Ribarič, Seismological Map of Slovenia for 500 Years Return Period, Seismological survey of Slovenia, Ljubljana, Slovenia, 1987.
15. J. Lapajne, B. Šket-Motnikar, and P. Zupančič, "Design ground acceleration map of Slovenia," Potresi V Letu 1999, pp. 40–49, 2001.
16. N. Ambraseys, P. Smit, R. Sigbjornsson, P. Suhadolc, and B. Margaris, "Internet-Site for European Strong-Motion Data," 2002, http://www.isesd.hi.is/ESD_Local/frameset.htm.
17. B. Šket-Motnikar and T. Prosen, "Accelerations in Poso je on July 12 2004," Potresi V Letu 2004, pp. 105–113, 2006.
18. S. Buser, Basic Geological Map of Yugoslavia 1: 100. 000—Sheet Tolmin and Videm, Geological Survey of Slovenia, Ljubljana, Slovenia, 1986.

19. M. Ribičič, R. Vidrih, and M. Godec, "Seismogeological and geotechnical conditions of buildings in Upper Soča Territory, Slovenia," Geologija, vol. 43, no. 1, pp. 115–143, 2000.
20. S. Buser, Basic Geological Map of Yugoslavia 1: 100. 000—Sheet Tolmin and Videm Explanatory Text, Geological Survey of Slovenia, Ljubljana, Slovenia, 1986.
21. A. Casagrande, "Classification and identification of soils," Transactions of the American Society of Civil Engineers, vol. 113, pp. 901–930, 1948.
22. D. Mayer-Rosa and M. J. Jimenez, Seismic Zoning: State-of-the-Art and Recommendations for Switzerland, Federal Office for Water and Geology, Swiss National Hydrogeological and Geological Survey, Berne, Switzerland, 2000.
23. A. Gosar, "Microtremor HVSR study for assessing site effects in the Bovec basin (NW Slovenia) related to 1998 Mw5.6 and 2004 Mw5.2 earthquakes," Engineering Geology, vol. 91, no. 2–4, pp. 178–193, 2007.
24. SIST, Slovenian Standard SIST EN, 1998-1 Eurocode 8: Design of Structures for Earthquake Resistance—Part 1: General Rules, Seismic Actions and Rules for Buildings—National Annex, Slovenian Institute for Standardization, Ljubljana, Slovenian, 2004.
25. CEN, Eurocode 8—Design of Structures For Earthquake Resistance, Part 5: Foundations, Retaining Structures and Geotechnical Aspects, European Standard, EN, 1998-5: 2004 (E), Stage 64, European Committee for Standardization, Brussels, Belgium, 2004.
26. M. Bavec, "New temporal, and genetic determinations of some late Quaternary sediments in the Bovec basin and its surroundings (NW Slovenia)," Geologija, vol. 45, no. 2, pp. 291–298, 2002.

Chapter 2

Remote Sensing and Geographic Information System for Fault Segments Mapping a Study from Taiz Area, Yemen

Anwar Abdullah, Shawki Nassr, and Abdoh Ghaleeb

Geology Department, Faculty of Applied Science, Taiz University, Taiz, Yemen

ABSTRACT

GIS and remote sensing data for allowing detection of structural features, such as faults, offer opportunities for improving mapping and identifying the areas that are likely to be locations of faulting areas. Landsat ETM-7 satellite data images were used and band-5 was found as the most suitable band for lineament delineation, based on the ability to identify geological features. Four contributing factors, namely, drainage patterns, faults (previously mapped), lineaments, and lithological contacts layers, were parameters used in this study to produce a fault potential prediction map using the overlay model techniques. The potential map (fault susceptibility map) classifies the study area into five potential zones, namely, very low, low, moderate, high, and very high potential. The areas covered by moderate to the highest potential zones were considered as fault segments (fault lines) in the area. The comparison of the potential map and the published fault map by using GIS matching techniques shows that 75 fault segments (fault lines) in the potential map were not properly identified in the study area. The correlation between fault segments and faults data collected from field work stations shows that there were 39 fault segments which may represent new faults in the area being identified. The presence of these faults is not known from the literature; this leads to updating and revising of existing geological map of the study area.

INTRODUCTION

Faults are weakness zones in the brittle part of the lithosphere, along which movement can take place in response to induced stresses. When faults undergo displacement, depending on geological and structural conditions, strain markers can be formed on the fault surface [1]. The presence of faults in any area is based on displacement of rock layers. But also, most of faults are represented by some geological features such as drainage patterns, lineaments (linear features), and lithological contacts between rock units within the rocks of the area. The presence of faults may be indicated by these geological features (factors).

The term lineament was first introduced by [2, 3] who recognized the existence of linear geomorphic features and interpreted them as surface expressions of zones of weakness or structural displacement of the earth's crust. Lineaments are linear features on the Earth's surface, usually related to the subsurface phenomena. Generally, lineaments are related to large fractures and faults where their orientation and number give an idea of fracture pattern of rocks [4]. In the recent years, the lineaments have been defined as natural crustal structures that may represent a zone of structural weakness [5].

The drainage system, which develops in an area, is strictly dependent on the slope, the nature, and attitude of bedrock and on the regional and local fracture pattern [6]. Most stream networks are adapted to regional slope and geological structures, picking out the main fractures in the underlying rocks [7].

The contact between two lithologies can also appear as a linear feature. This contact may appear as a change in drainage pattern across the structural features [8] or the two units may have different spectral properties [9].

Based on the definitions of lineaments, drainage patterns (including pattern the length, spatial distributions), and the lithological contacts between different rock types, the faults could be mapped in the study area. These geological features mostly resemble a fault lines in the area. The most important features in the area are the presence of drainage lines patterns and fractures. Lineaments, drainages, lithological contacts, and previous fault lines data are important data and used in this research for fault segments mapping using GIS technique.

With the advent of remote sensing and computer technology in the geosciences, geological investigation and interpretation have entered a new era. Remote sensing technology is very efficient for collecting data. Computer technology, such as computer-based geographic information system (GIS), supplies a different method for data storage, integration, analysis, and display. The combination of remote sensing and GIS provides an optimum system for various geological investigations such as fault mapping [10].

Several studies have been carried out to enhance geological knowledge and revise existing geological maps by using optical remotely sensed data and discovered new faults that were previously unmapped [11–16]. Furthermore, authors in [17, 18] conclude by suggesting that satellite images and/or geographic information systems (GIS) are useful for tectonic mapping and identifying previously unmapped faults.

The benefit of integrating different data and techniques in structural features such as faults identification and analysis made possible using remote sensing and GIS techniques is the ability to identify faults based on their characteristics. Creation and classification of various types of geological features can be carried based on the purpose of the study. Once the identification, preparation, and processes are complete, geographic information systems (GIS) functionality, such as vector and raster spatial analysis and overlay, can be employed for structural mapping and analysis using powerful software programs [19].

Generally, Yemen is located in the southwestern part of Arabian Peninsula, and it is affected by an active rifting zone of the Gulf of Eden and the Red Sea opening. This active (rifting) zone can cause the reactivation of the old fault systems and may create new faulting lines in the area. The active faults (such as, rock mass movements or sliding on the fault planes and bedrock lithologies cracking) may cause severe damages in some places, especially in urban or settlements areas. For example, bedrock cracking and rock mass sliding happened in the 2003 and 2010 in some places in Taiz city caused severe damages, and many houses, roads, and other infrastructures were destroyed.

The previous works (data) such as geological and structural maps are very old data, showing a few of fault lines in the study area. For this purpose, mapping of faults may lead to updating these data. In this paper, the using of remote sensing (data processing) and GIS (modeling) techniques could be helpful tools for mapping of faults, where the different geological features (factors) are taken in the considerations. The results of the work could be considered from the government or the people before or during the different construction projects, to increase the safety, especially in the urban areas.

The purpose of this study was to test the integration between remote sensing and geographic information system (GIS) for detecting the fault segments (fault lines) over the study area and to investigate the ability of this method in giving real results compared to the previous data with respect to the ground truth (field work).

The study area located in the southwest of Yemen has an area of 176 km^2. It includes Taiz city and extends between Rubayi and Al-Hawban areas (Figure 1).

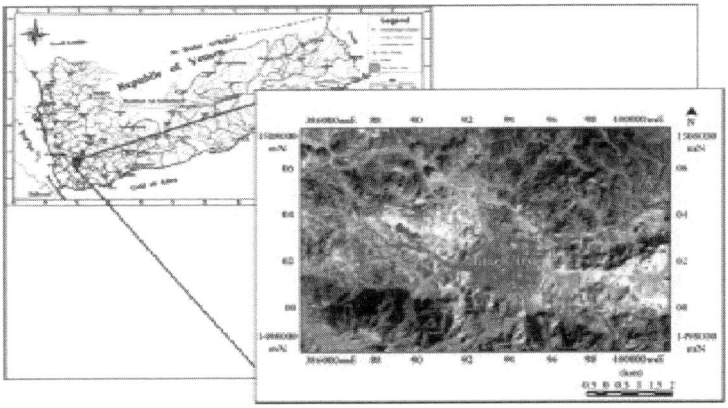

Figure 1: Location of the study area.

According to GSY [20] classification, the rocks in the study area are divided into two groups as follows.

- Tertiary (granite body intrusive)
- Tertiary (volcanic rocks such as basalt, rhyolite, dacite, ignimbrite, and ashflow deposits).

The geological map of the study area (Figure 2) was digitized from the geological sheet map of Taiz with scale 1 : 250,000.

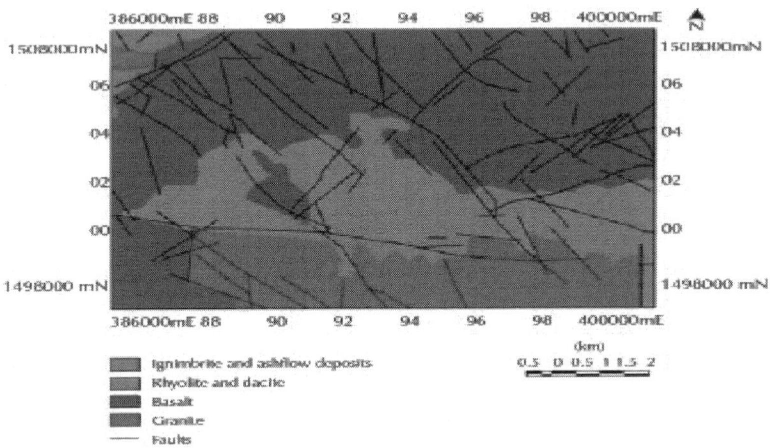

Figure 2: Geological map of the study area [20].

MATERIALS AND METHODS

Various data were used for fault segment susceptibility mapping (potential fault zones mapping) in the study area. Various image processing and enhancement techniques were applied on different remote sensing data including the Landsat ETM-7 (Enhanced Thematic Mapper) satellite images and SRTM (The Shuttle Radar Topography Mission) to get maximum information by using different approaches of extraction methods. The other data used in this study were topographic maps, geological maps, and field data.

During conducting this study, many different software packages were used since there was no single software that would process all steps in the analyses. Therefore, the diverse software used for the analysis in the current study is PCI Geomatica (version 9.1), ERDAS (version 8.4), and Ilwis software (version 3.3).

Data Processing

According to this paper the linear features delineation was based on decision of the most appropriate data, such as drainage patterns, lineament, fault (previously mapped), and lithological contact layers for mapping of fault segments using GIS. Many of researchers such as [21–26] have utilized lineaments, drainage patterns, faults, and lithology factors as important factors to measure and map the susceptibility of geological hazard. Therefore, the input attributes of the structural features of an area should be studied as a first consideration.

The previous data about the study area including geological, structural, and topographic maps are digitized. To prepare these maps, first, the maps are converted into a digital format by using the scanner. There are some data which were extracted from data in digital formats, such as Landsat ETM satellite images and SRTM.

Generally, four digital layers (maps) such as drainage patterns, fault (previously mapped), lineament, and lithological contact were prepared and converted into secondary data such as drainage buffer, fault buffer, lineament buffer, and lithological contacts buffer layers.

These layers were converted into slicing layers with five classes. Each class was determined as 30 m. This is because these layers were generated from different data (ETM, SRTM, and previous maps) with different spatial resolutions and scales. Due to this, these layers should be enhanced and resampled (rescaled) to be suitable in resolution for GIS model. The resolution was selected as 30 m for this process based on the ability of satellite image of 30 m spatial resolution to identify the geological features in the area. Then, slicing layers were converted into weight layers (thematic layers). These procedures are illustrated in Figure 3.

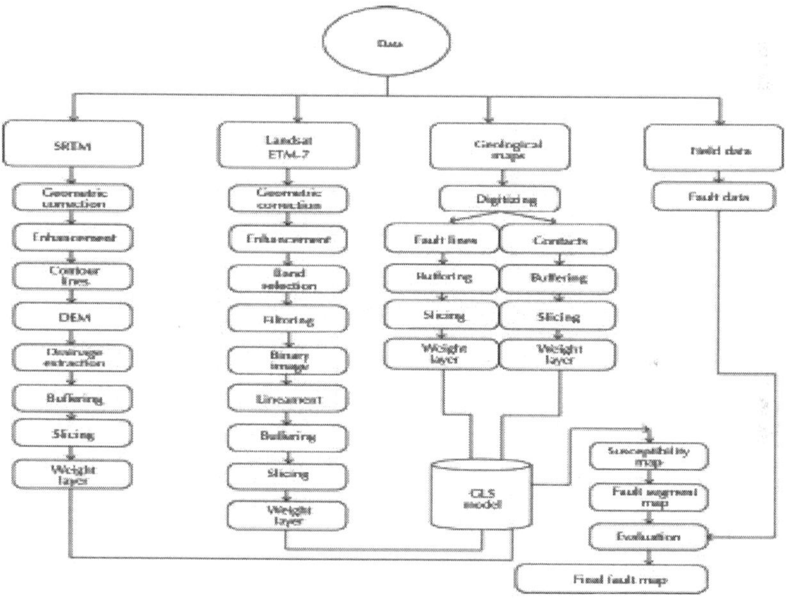

Figure 3: Chart of the methodology in this study.

Arbitrary classifications are still common; however, the main classification approaches are ranking, natural breaks, equal interval classes, equal area classes, and mean value and standard deviation intervals [25, 27–29]. Hence, the classification determines the spatial distribution of buffering zones or susceptibility (classifying susceptibility) based on equal area classes.

Classes (five classes of each weight layer) and their weight values are given in Table 1. Each class has a weight to express the contribution to the occurrence of fault segments. The weight value was assigned to be between 1 to 5 (Table 1). These weight layers (maps) were input into GIS model. The resultant map (susceptibility map) was classified into different zones, very low, low, moderate, high, and very high fault susceptibility zones or equal area classes using ranking approach.

Table 1: Parameters and the weight values are given to different factors in the study area

Classes	Weight value
<30 m	5
30–60 m	4
60–90 m	3
90–120 m	2
>120 m	1

Lineament Factor

Lineaments are related to large structural fractures [4] and it may represent zones of weakness [22]. Also, the presence of tectonic structures such as linear features could be important factor in geological hazard occurrences [30, 31].

Filtering is useful in image enhancement. An edge-enhancing filter can be used to highlight any changes of gradient within the image features, such as structural lines [17, 28, 32–37]. They generated lineament maps and determined several significant structural features.

In the recent years, the lineaments have been defined as natural crustal structures that may represent a zone of structural weakness [38]. Landsat ETM-7 satellite data were used and the first step was to select the band that should be used for lineament extraction [39]. Visual inspection of the individual bands was carried out, based on the ability to identify features, and band-5 (1.55–1.75 μm) (SWIR) was selected as shown in Figure 4. And it was stretched linearly to output range from 0 to 255 (Figure 5), because it is the least affected band by the scattering, through the travel path of the long wavelength in the atmosphere. Therefore, it shows a good contrast and better display of geological features compared to other bands.

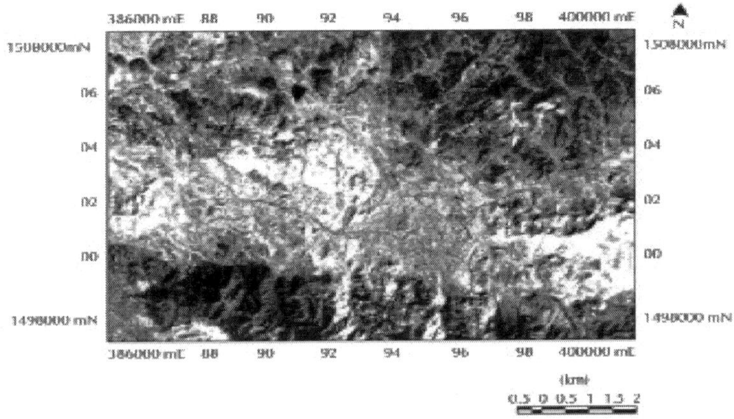

Figure 4: Original band-5 of Landsat ETM-7.

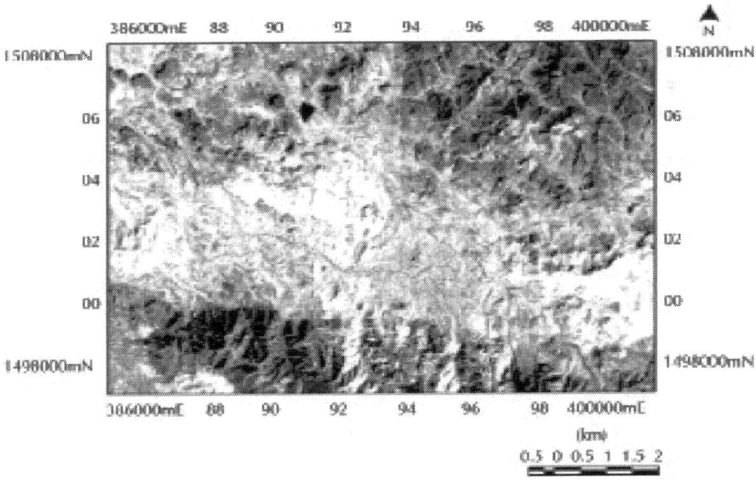

Figure 5: Band-5 of Landsat ETM-7 after contrast enhancement.

The second step was to select the filter type. For this purpose, different types of filters are tested such as 3 by 3 sobel kernel filter, 5 by 5 edge kernel filter, and 7 by 7 edge kernel filter. The 5 by 5 and 7 by 7 edge enhancement filters give thicker and less linear areas after threshold application; sobel results in thinner and more linear zones that give the best result compared with other two filters.

The sobel filter was the best which convolved over band-5 in four principal directions and the filtered image was added back to the original image. The sobel filters in four principal directions are given in Table 2.

Table 2: Sobel kernel filters in four main directions

Four main principal directions											
NW-SE			E-W			NW-SE			N-S		
−2	−1	0	−1	−2	−1	0	1	2	1	0	−1
−1	0	1	0	0	0	−1	0	1	2	0	−2
0	1	2	1	2	1	−2	−1	0	1	0	−1

Then, the four enhanced filtered images combined together to produce one threshold filtered image (Figure 6), and this image was converted to binary image by applying adaptive thresholding. This adaptive threshold used to create binary images is controlled by two parameters, and the suitable values of adaptive thresholding parameters are windows size of 3, and this value is equal to thinning of pixels and the mean offset of 3. The final binary image of the study area is shown in Figure 7.

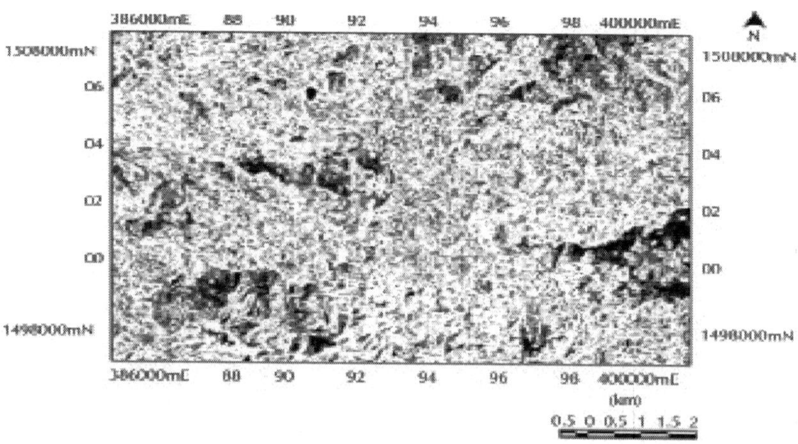

Figure 6: Threshold image produced by combining four filtered images.

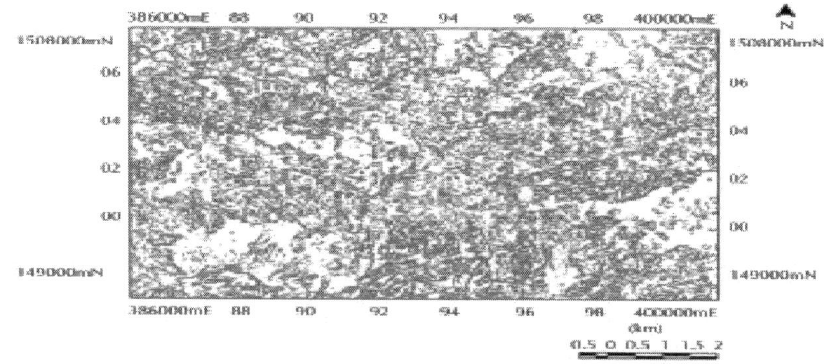

Figure 7: Binary image produced by using sobel filters.

The lineaments detected during the interpretation process were digitized directly on the image on the screen, and the final lineament map was recorded and stored in vector files (Figure 8).

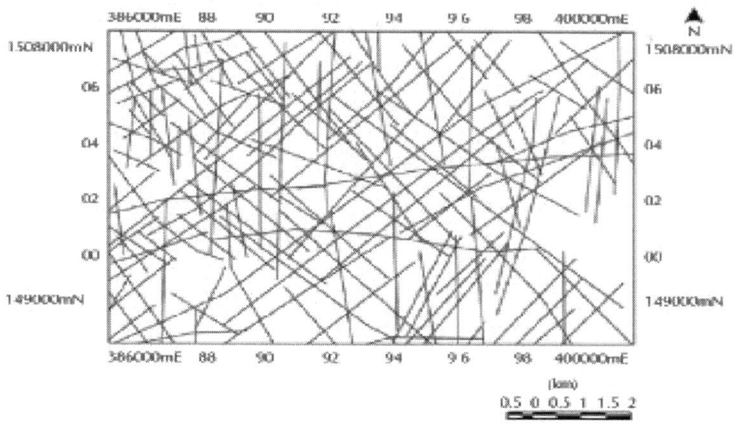

Figure 8: Lineaments extracted from binary image.

There are important points which need to be mentioned and should be followed in relation to the above procedure for lineament mapping. First, care is needed in the interpretation of the result of such analysis to exclude man-made features; therefore, the final map was screened to remove lineaments related to cultural features such as roads, canals, and field plantation boundaries by comparing it with the topographic maps and geological map of the area. All procedures mentioned above have been followed during trace lineaments from binary image.

The lineament map was prepared and converted into lineament buffer layer and then converted into slicing layer with five classes. Finally, this layer was converted into weights layer (Figure 9). The weight value was assigned to be between 1 and 5 (Table 2). And, this layer was ready to be input into GIS modeling.

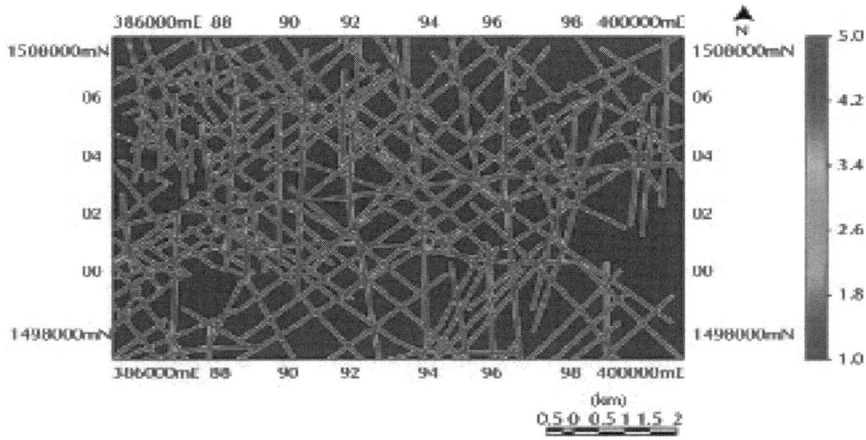

Figure 9: Lineament weight layer.

Drainage Patterns Factor

The drainage pattern is apparently being controlled by structure and lithology in the study area. The lithologic variation has given a rise to different drainage patterns. For example, radial and dendritic drainage patterns are developed over granitic rocks. Moreover, the most important feature in the area is the presence of drainage lines patterns and fault lines. It is clearly to see that there is a good relationship between these fault lines and drainage pattern system distribution especially with third and fourth river orders in the area.

Geological features are any alignment of features on satellite images such as the various types recognized including topographic, drainage, vegetative, and color alignments. Digital elevation models

(DEMs) are very useful in aiding the classification of landforms for many purposes such as recognition of drainage patterns. Extraction of topographic feature information from DEMs has become increasingly popular in structural analysis [40]. Digital elevation models (DEMs) data were used to trace tectonic features and mapping geologically and topographically defined structures in many areas [41]. The Shuttle Radar Topography Mission (SRTM) data were also used to trace drainage pattern system and mapping structures features [42]. The extraction of negative system lines (fault lines) from DEM data mostly resembles drainage lines patterns [43].

SRTM radar images with 90 m resolution were used to obtain contour line data. This contour data was used to create the DEM with 30 m resolution of the study area. The interpolation and resampling processing techniques were used to convert the contour line data (polygon) into 30 m grid layer and the grid layer was converted into a DEM with 30 m spatial resolutions (Figure 10).

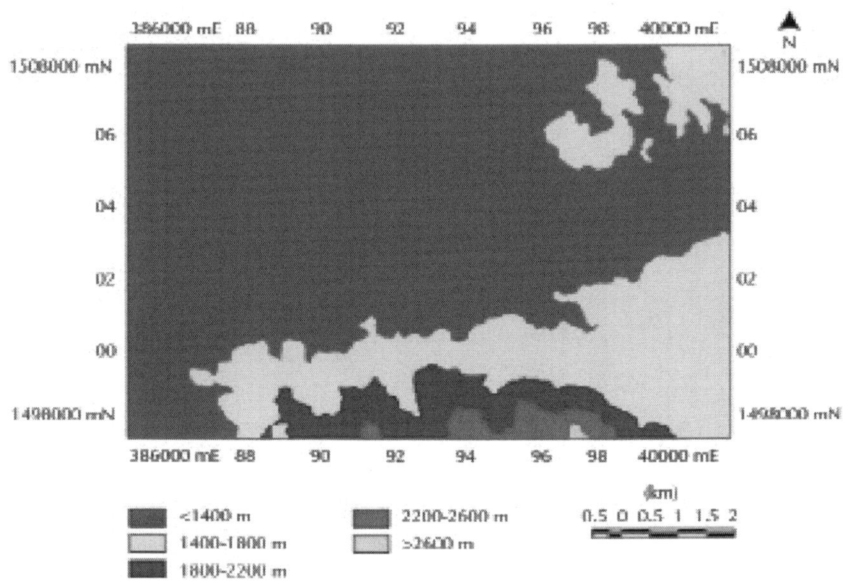

Figure 10: Digital elevation model (DEM) of the study area.

This DEMs data was used for automatic extraction of drainage patterns over the study area using drainage basin from elevation data function (DRAIN) of PCI Geomatica software version 9.1. The drainage basin from elevation data function (DRAIN) locates the drainage networks (river or stream) within the DEM. The output data channel will appear as lines representing the rivers and streams [44].

The result of the drainage pattern map using automatic extraction technique is shown in Figure 11. This technique helps us to extract the drainage patterns over the settlement area, and the topographic maps were used as reference data.

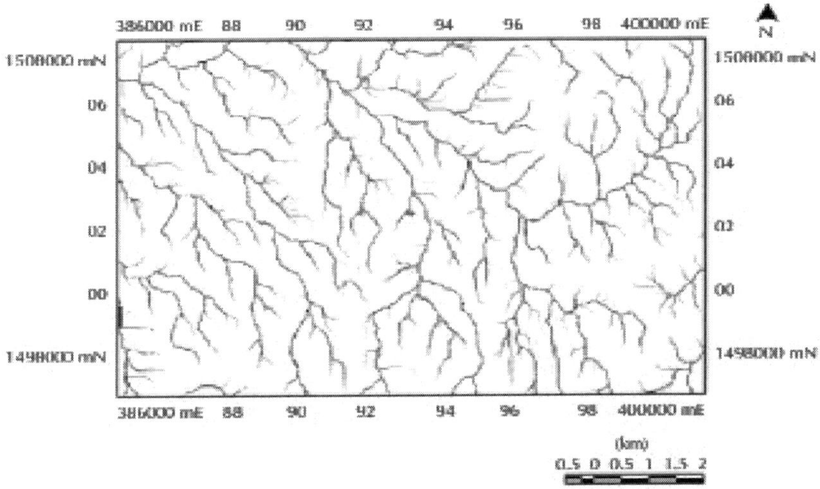

Figure 11: Drainage patterns map showing the nature of drainage system in the area.

This map was converted into vector layer. The drainage pattern layer (vector layer) was converted into drainage buffer layer and then converted into slicing layer with five classes. This map was converted into weights layer (Figure 12). The weight value was assigned to be between 1 and 5 (Table 1). This layer was ready to be input into GIS modeling.

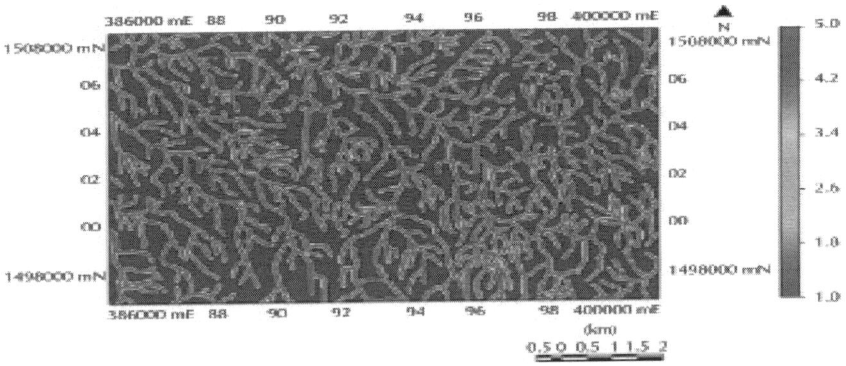

Figure 12: Drainage weight layer.

Lithological Contact Factor

Lithological contacts were found in the field, such as bedding planes and lithological contacts between different rock types. The lithological contact layer of the study area (Figure 13) was digitized from the geological sheet map of Taiz with scale 1 : 250,000. The lithological contact layer (vector layer) was converted into lithological buffer layer and then converted into slicing layer with five classes. And this map was converted into weights layer (Figure 14). The weight value was assigned to be between 1 and 5 (Table 1). This map was ready to be input into GIS modeling.

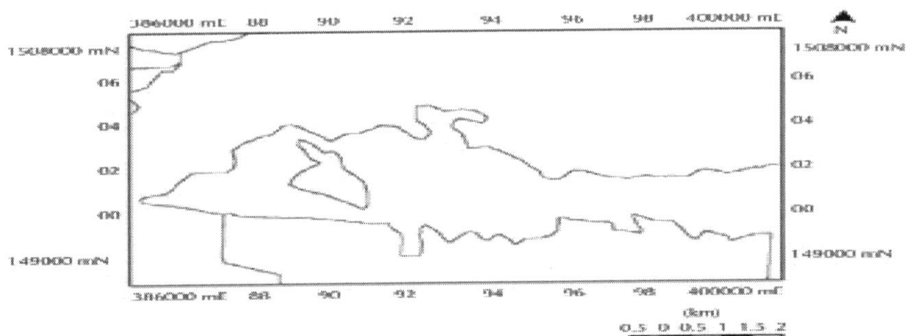

Figure 13: Lithological contacts layer [20].

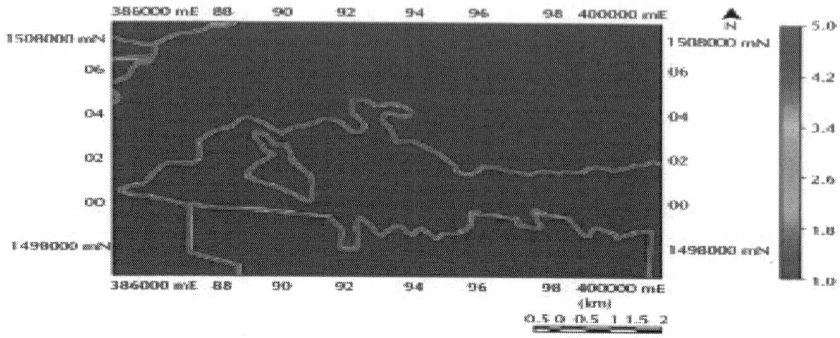

Figure 14: Lithological contacts weight layer.

Fault Factor

Geological structures are patterns such as faults, folds, joints, and others found in the rocks resulting from natural deformation [45]. The fault map of the study area (Figure 15) was digitized from the geological and the hydrogeological sheet maps of Taiz area with scales 1 : 250,000 and 1 : 50,000. The first sheet was prepared by geological survey of Yemen (GSY) and the second was prepared by National Water Resources Authority, Dar El-Yemen Hydro consultant [46]. This layer (fault layer) was converted into fault buffer layer and then converted into slicing layer with five classes. And this map was converted into weights layer (Figure 16). The weight value was assigned to be between 1 and 5 (Table 1). This layer was ready to be input into GIS modeling.

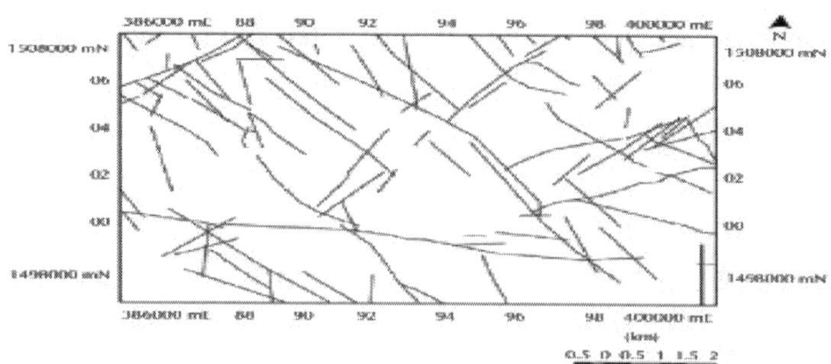

Figure 15: Fault map of the study area [20, 46].

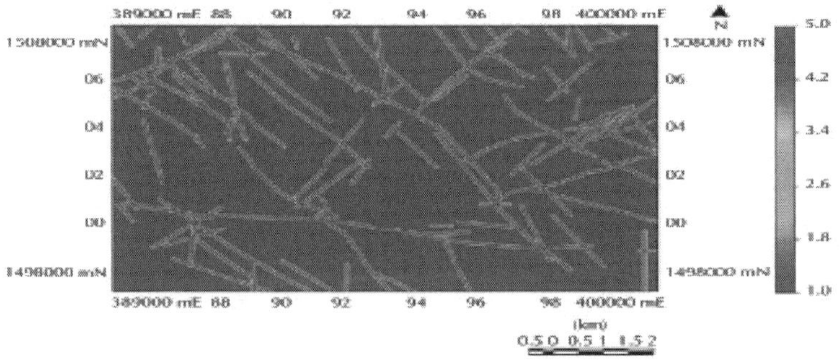

Figure 16: Fault weight layer.

GIS Model

The major goal of this study is to produce the fault susceptibility and the fault segment maps in order to get information about the possible locations of new faults existing in the area.

Generally, four digital weight layers such as drainage patterns, fault (previously mapped), lineament, and lithological contact layers were prepared from different data. The first step of the GIS modeling involves the input data (weight layers), and the second step includes selection of the suitable equation for this model to determine the susceptibility of faulting segments in the study area.

In this study, the overlay layers model (spatial multilayers analysis) was used preformed on these thematic maps with different weights by combining and aggregating these maps together with considering all different parameters. The fault susceptibility model produced with a modified linear equation of drastic model [47], using the following equation:

$$\text{Potential Map} = DRN_W + FLT_W + LIN_W + GCT_W, \quad (1)$$

where, DRN_W is the drainage weight layer, FLT_W the fault weight layer (previously mapped), LIN_W the weight layer, and LCT_W the lithological contact weight layer.

Researchers such as [48–50] classified the susceptibility into five potential classes, very low (VL), low (L), medium (M), high (H), and very high (VH).

The resultant map (susceptibility map) was classified into different zones, including very low, low, moderate, high, and very high fault susceptibility zones or equal area classes using ranking values.

Evaluation of the Susceptibility Map

The above-mentioned technique was used to produce the fault segment susceptibility map from the different data. The result obtained from this technique needs to be evaluated in some manner. The first evaluation was made with respect to the fault map (previous fault map) using GIS overlay technique to determine where the fault segments and true faults are matched over the study area.

During fieldwork, structural data (strike/dip of faults) were collected from 75 stations in the study area. These data were used for the fault segment map evaluation. These stations were located in the fault segments map to make whether there is any correlation between the fault segments and the faults collected from the field.

RESULTS AND DISCUSSION

The final susceptibility map of faulting (potential fault zones) was produced as a result of the integration of all parameters of weight layers (drainage, fault (previously mapped), lineament, and lithological contact). This map was classified into five zones, namely, very low potential zone (VL), low potential zone (L), medium potential zone (M), high potential zone (H), and very high potential zone (VH), as shown in Figure 17.

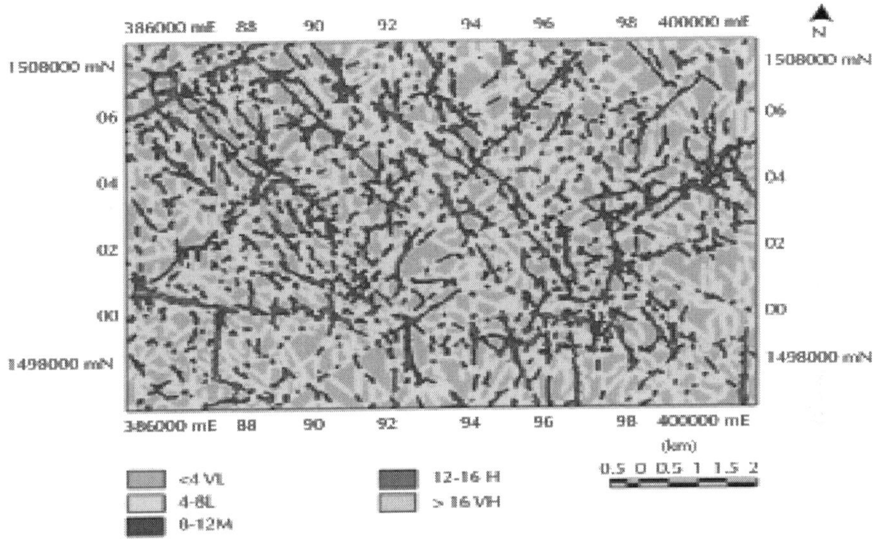

Figure 17: Susceptibility map of fault zones over the study area, classified into five zones, namely, very low potential zone (green), low potential zone (yellow), medium potential zone (blue), high potential zone (pink), and very high potential zone (cyan).

The potential zones in the fault susceptibility map show that the very low potential zone has a total area of 22.87%, and the low potential zone has a total area of 51.89%. More than half of the study area are dominated by very low and low potential zones. And this area does not reflect faults (faults in the previous map), while the 22.40% of moderate, 2.73% of high, and 0.11% of very high potential zones areas almost cover all the faults.

The evaluation of the susceptibility map has been computed by using GIS overlay capability techniques with the fault map of the study area to determine where the potential zones and faults are matched. In order to evaluate our susceptibility map for faulting, a comparison with the known structural fault lines has been done. The result of this technique shows that all of the faults were matched with the moderate to high and very high potential zones (continuous areas from moderate to high and very high).

There were some of moderate to high and very high potential areas in the susceptibility map that could not match any fault lines in the output map of the fault map. These fault segments identified by this technique may be surface reflections of the new fault locations in the study area which were not identified by previous works. This means that the use of this applied model worked successfully to identify linear features (e.g., fault lines) in the area of interest. The results suggest an additional 75 of new fault segment lines, which were not recognized by the fault map over the study area (Figure 18).

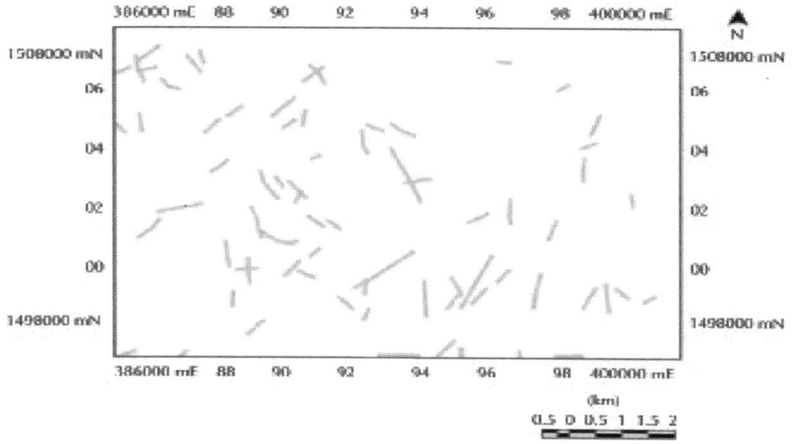

Figure 18: The 75 fault segments (faults lines) as a vector layer extracted from the susceptibility map over the study area.

Moreover, the fault segments obtained from this model were shorter in length compared with faults that were observed in the field. For this reason, this paper attempted to determine the accuracy of this technique for fault segments mapping and the relationship between these segments and ground truth (faults) in terms of the length and the spatial distribution in the study area.

Based on the pattern of fault segments including the length, spatial distributions, and visual interpretation with a topographic map, a drainage pattern map, and three dimensions of DEM and Landsat ETM, the true length of fault segments could be recognized from this analysis and could be implemented in order to assess the extraction accuracy of every true fault length in the area.

Validation of result was made with the geological truth (field work) in the study area. During fieldwork, many faults were identified in different field work outcrops. Strike/dip readings of faults have been collected during the field work from different rock outcrops in the area. Based on the field data, these linear features identified on Landsat imagery and DEM are interpreted to be faults (Figure 19).

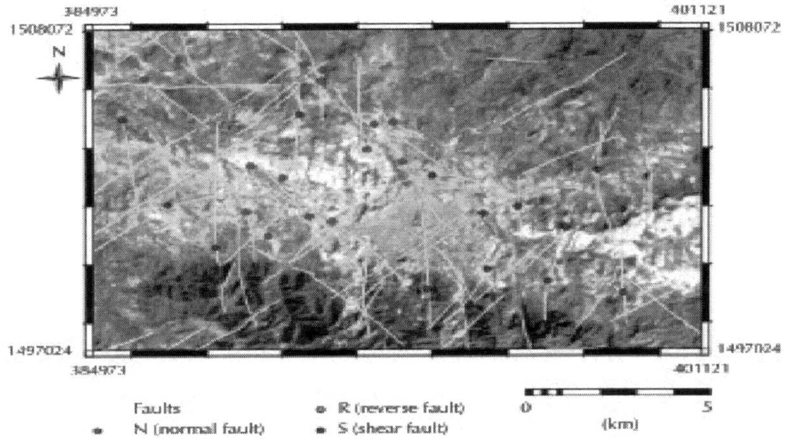

Figure 19: The 39 fault segments identified as faults in the study area showing true fault length and points of field work stations, normal fault segments indicated with red points, reverse fault segments indicated with green points, and strike slip fault segments indicated with blue points.

The fieldwork stations were located in the fault segments map to make whether there is any correlation between the fault segments (faults lines identified by this technique) and the faults data collected from field work stations. There was good correlation between a few fault segments lines and faults, where 39 locations came in the same or nearby the location of fault segments (faults lines). Therefore, these 39 fault segments (faults lines) were identified as new faults (Figure 19). The structural reading (strike/dip) of faults from 93 fieldwork stations are given in Table 3.

Table 3: Structure reading of faults from the field

Location	North	East	Strike/ dip	Name of fault	Location	North	East	Name of fault
1	1506 721	387 493	N224E /43°	Normal	21	1504 706	392729	Normal

2	1505 753	387 024	N223E /56°	Normal	22	1504 886	392998	Normal
3	1507 429	386 266	N308E /78°	Reverse	23	1502 642	386915	Reverse
4	1503 679	386 486	N311E /16°	Reverse	24	1501 106	385718	Reverse
5	1505 813	389 149	N312E /79°	Reverse	25	1499 311	395042	Shear
6	1500 837	396 967	N329E /58°	Shear	26	1500 278	390156	Reverse
7	1503 011	398 523	N184E /93°	Shear	27	1500 228	392968	Reverse
8	1505 863	390 066	N313E /96°	Reverse	28	1503 100	393357	Normal
9	1499 101	398 982	N4E /12°	Shear	29	1502 003	388032	Normal
10	1507 229	390 136	N228E /66°	Normal	30	1505 354	390315	Shear
11	1505 334	389 717	N315E /8°	Reverse	31	1502 412	388301	Shear
12	1506 551	390 625	N15E / 65°	Normal	32	1499 999	393786	Shear
13	150 2303	389 926	N219E /45°	Normal	33	1500 358	393906	Shear
14	1501 624	390 674	N225E /15°	Normal	34	1502 293	394434	Shear
15	1500 837	391 781	N226E /11°	Normal	35	1498 832	398294	Reverse
16	150 2632	390 305	N227E /24°	Normal	36	1503 579	385947	Shear
17	15043 97	389 448	N305E /18°	Reverse	37	1502 143	387334	Shear
18	1499 061	397 037	N352E /47°	Shear	38	1504 347	398483	Reverse
19	1501 994	396 239	N360E /50°	Shear	39	1505 903	397376	Reverse
20	150 2861	389 757	N217E /47°	Normal				

Three fault systems were identified. These are the E-W, NW-SE, and NE-SW fault systems [46]. The area was subjected to intensive tectonic activity which resulted in a series of grabens, horsts, and tilted fault blocks. The tectonic activity deformed the area before the Tertiary volcanic activity, especially through the NE-SW faults which run parallel to the rift system of the Gulf of Aden, and it is considered the oldest fractures in the area. The NW-SE faults are run parallel to the opening of the Red Sea and occurred in the area during the Tertiary volcanic activity. The N-S and E-W fault system are the youngest major faults in the area. Smaller faults (as well as joints) also follow the same trends as the major faults. However, more intensive faulting is in the NW-SF direction [46].

Conjugate faults completely defined the principal stress direction. Strike and dip of the faults were plotted in equal-area projection of Schmidt net to obtain and calculate the directions of principal stresses which acted on those faults.

Totally 10 strike-dip readings of the conjugate faults from different rock units of the area were analyzed with the Stereonet program (Stero32 Version 0.9). The stereographic plot was prepared (Figure 20). According to the result of the conjugate faults analysis, compressional principal stress (σ1) is measured to be in the NE-SW direction.

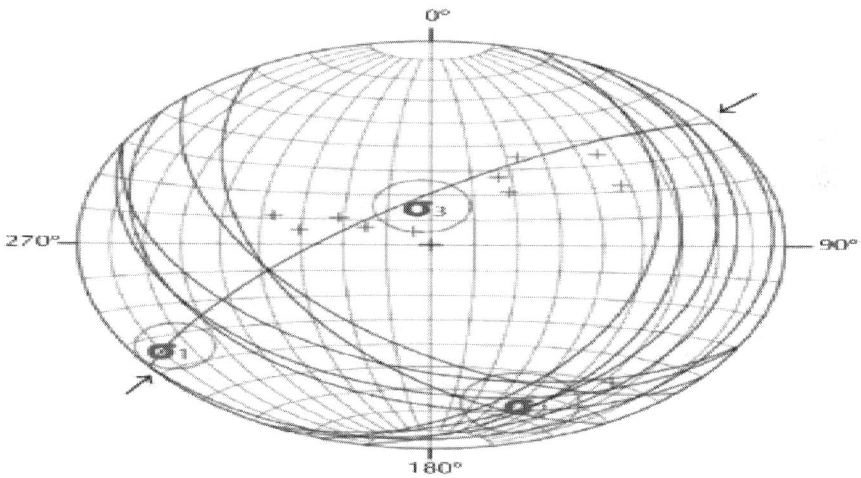

Figure 20: Result of conjugate faults using stereographic analysis.

This analysis indicates that the N-S and E-W orientated fault sets are sets of shear fractures which could be typical for strike slip faults. The NW-SE and NE-SW orientated sets are sets of tensional and extensional fractures, which could be typical for normal and reverse faults, respectively. Different field observations have also indicated that many damages were found or occurred along the NW-SE faults.

CONCLUSIONS

This study shows that remote sensing data and GIS technique are powerful tools in identifying geological structures such as fault segments in the urban area such as the Taiz city, Yemen. The use of different types of data such as drainage patterns, faults (previously mapped), lineament, and lithological contacts with spatial resolution of 30 m was combined through a process of integration and index overlay modeling technique for producing the susceptibility map (potential fault zones) of fault segments in the study area. GIS spatial overlay technique was used to determine the spatial relationships of all the criteria (factors) and subcriteria (classes) within layers (maps) to classify and map the potential area of fault zones of the study area. The fault segments map shows that there is good correlation and distribution between fault segments and faults (previous fault lines) in the study area. There are 39 fault segments which may represent new faults in the area being identified. The presence of these faults is not known from the literature. These faults could be active due to the rift system of the Gulf of Eden and opening Red Sea. Active or reactivated faults can cause enormous damage to settlement or urban (houses or construction project) areas by rock mass movement along the fault surfaces or bedrock cracks, damage which refers to loss of life or injury, or damage to roads. Furthermore, geological survey of Yemen is requested to update the maps and create new ones using new data and techniques and advise the people about the proper sitting of houses and restricting settlement activities in danger areas. GIS models can then yield more accurate estimates of the fault mapping using more data such as geophysical data.

ACKNOWLEDGMENT

The authors would like to thank the Geology Department, Faculty of Applied Science, Taiz University, Taiz, Yemen for the support to this research and staff for their valuable helps and comments.

REFERENCES

1. B. Dehandschutter, Study of the structural evolution of continental basins in Altai, central Asia [Ph.D. thesis], 2001.
2. W. H. Hobbs, "Lineaments of the Atlantic Border region," Geological Society, vol. 15, pp. 483–506, 1904.
3. W. H. Hobbs, "Repeating patterns in the relief and in the structure of the land," Geological Society, vol. 22, pp. 123–176, 1911.
4. L. E. Arlegui and M. A. Soriano, "Characterizing lineaments from satellite images and field studies in the central Ebro basin (NE Spain)," International Journal of Remote Sensing, vol. 19, no. 16, pp. 3169–3185, 1998.
5. R. T. Walker, "A remote sensing study of active folding and faulting in southern Kerman province, southeast Iran," Journal of Structural Geology, vol. 28, no. 4, pp. 654–668, 2006.
6. C. Travaglia and N. Dainelli, Groundwater Search by Remote Sensing: A Methodological Approach, Environment and Natural Resources, FAO, Rome, Italy, 2003.
7. M. Morisawa, Rivers, Longman, New York, NY, USA, 1985.
8. C. E. Brockmann, A. Fernandez, R. Ballon, and I. I. Claure, "Analysis of geological structures based on landsat-1 images," in Remote Sensing Applications for Mineral Exploration, W. L. Smith, Ed., pp. 292–317, Dowden, Hutchinson and Ross, Strondsberg, Pa, USA, 1977.
9. P. T. Nguyen and D. Ho, "Multiple source data processing in remote sensing," in Digital Image Processing in Remote Sensing, J. P. Muller, Ed., pp. 153–176, Taylor and Francis, Philadelphia, Pa, USA, 1988.
10. X. Chen, Application of remote sensing and GIS techniques for environmental geologic investigation, northeast Iowa [Ph.D. thesis], University of Iowa, Iowa, Iowa, USA, 1992.
11. A. Ü. Akman and K. Tüfekçi, "Determination and characterization of fault systems and geomorphological features by RS and GIS techniques in the WSW part of Turkey," in Proceedings of the 20th ISPRS Congress, pp. 899–904, Istanbul, Turkey, 2004.
12. P. R. E. Guerra, Faulting evidence of isostatic uplift in the rincon mountains metamorphic core complex, an image processing analysis [Ph.D. thesis], 2000.

13. M. A. Juhari and A. Ibrahim, "Geological applications of Landsat TM imagery: mapping and analysis of lineaments in NW Peninsula Malaysia," in Proceedings of the 18th Asian Conference on Remote Sensing, pp. J-1-1–J-1-8, Kuala Lumpur, Malaysia, 1997.
14. W. F. P. G. Micheal, "Lineaments analysis South Florida region, aquifer storage and recovery regional study," Draft Technical Memorandum, Central and Southren Florida Project, Army Corps of Engineers, Jacksonville, Fla, USA, 2004.
15. I. D. Novak and N. Soulakellis, "Identifying geomorphic features using LANDSAT-5/TM data processing techniques on Lesvos, Greece," Geomorphology, vol. 34, no. 1-2, pp. 101–109, 2000.
16. S. Solomon and W. Ghebreab, "Lineament characterization and their tectonic significance using Landsat TM data and field studies in the central highlands of Eritrea," Journal of African Earth Sciences, vol. 46, no. 4, pp. 371–378, 2006.
17. K. S. Kavak and H. Cetin, "A detailed geologic lineament analysis using landsat TM data of Gölmarmara/Manisa region, Turkey," Online Journal of Earth Sciences, vol. 1, no. 3, pp. 145–153, 2007.
18. L. A. Rutty, The basement fracture pattern of sothern Ontario: a tectonic interpretation based on landsat TM imagery, airphotos and field data [M.S. thesis], National Library of Canada, Ottawa, Canada, 1993.
19. K. M. M. Elias, "Multiple data set integration for structural and stratigraphic analysis of Oil and Gas bearing formation using GIS," in Proceedings of the Map India Conference, Geology & Mineral Resource, 2003.
20. Geological Survey of Yemen (GSY), Geological Map of Taiz Area (1:250, 000), 1990.
21. S. Bai, J. Wang, G. N. Lu, P. G. Zhou, S. S. Hou, and S. N. Xu, "GIS-based logistic regression for landslide susceptibility mapping of the zhongxian segment in the three gorges area, China," Geomorphology, vol. 115, pp. 23–31, 2010.
22. R. L. Bates and J. A. Jackson, Glossary of Geology, American Geological Institute, Alexandria, Va, USA, 1987.
23. M. L. Süzen and V. Doyuran, "Data driven bivariate landslide susceptibility assessment using geographical information

systems: a method and application to Asarsuyu catchment, Turkey,"Engineering Geology, vol. 71, no. 3-4, pp. 303–321, 2004.

24. F. Guzzetti, A. Carrara, M. Cardinali, and P. Reichenbach, "Landslide hazard evaluation: a review of current techniques and their application in a multi-scale study, Central Italy," Geomorphology, vol. 31, no. 1–4, pp. 181–216, 1999.
25. C. F. Chung and A. G. Fabbri, "Probabilistic prediction models for landslide hazard mapping,"Photogrammetric Engineering and Remote Sensing, vol. 65, no. 12, pp. 1389–1399, 1999.
26. F. C. Dai and C. F. Lee, "Landslide characteristics and slope instability modeling using GIS, Lantau Island, Hong Kong," Geomorphology, vol. 42, no. 3-4, pp. 213–228, 2002.
27. L. Ayalew, H. Yamagishi, and N. Ugawa, "Landslide susceptibility mapping using GIS-based weighted linear combination, the case in Tsugawa area of Agano river, Niigata prefecture, Japan," Landslides, vol. 1, pp. 73–81, 2004.
28. T. B. Minor, J. A. Carter, M. M. Chesley, and R. B. Knowles, "An integrated approach to groundwater exploration in developing countries using GIS and remote sensing," in Proceedings of the International American Congress on Surveying and Mapping/ American Society for Photogrammetry and Remote Sensing (ACSM/ASPRS ‹94), pp. 418–428, 1994.
29. S. Lee and K. Min, "Statistical analysis of landslide susceptibility at Yongin, Korea," Environmental Geology, vol. 40, no. 9, pp. 1095–1113, 2001.
30. L. Donati and M. C. Turrini, "An objective method to rank the importance of the factors predisposing to landslides with the GIS methodology: application to an area of the Apennines (Valnerina, Perugia, Italy)," Engineering Geology, vol. 63, no. 3-4, pp. 277–289, 2002.
31. A. Günther, "SLOPEMAP: programs for automated mapping of geometrical and kinematical properties of hard rock hill slopes," Computers and Geosciences, vol. 29, no. 7, pp. 865–875, 2003.
32. S. A. Ali and S. Pirasteh, "Geological applications of Landsat Enhanced Thematic Mapper (ETM) data and Geographic Information System (GIS): mapping and structural interpretation

in south-west Iran, Zagros Structural Belt," International Journal of Remote Sensing, vol. 25, no. 21, pp. 4715–4727, 2004.

33. G. Jordan, B. M. L. Meijninger, D. J. J. V. Hinsbergen, J. E. Meulenkamp, and P. M. V. Dijk, "Extraction of morphotectonic features from DEMs: development and applications for study areas in Hungary and NW Greece," International Journal of Applied Earth Observation and Geoinformation, vol. 7, no. 3, pp. 163–182, 2005.
34. A. Anwar, M. A. Juhari, and A. Ibrahim, "The extraction of lineaments using slope image derived from digital elevation model: case study of Sungai Lembing—Maran area, Malaysia," Journal of Applied Sciences Research, vol. 6, no. 11, pp. 1745–1751, 2010.
35. K. Koike, S. Nagano, and M. Ohmi, "Lineament analysis of satellite images using a Segment Tracing Algorithm (STA)," Computers and Geosciences, vol. 21, no. 9, pp. 1091–1104, 1995.
36. A. Mah, G. R. Taylor, P. Lennox, and L. Balia, "Lineament analysis of Landsat Thematic Mapper images, Northern Territory, Australia," Photogrammetric Engineering & Remote Sensing, vol. 61, no. 6, pp. 761–773, 1995.
37. A. Anwar, M. A. Juhari, and A. Ibrahim, "A comparison of landsat TM and SPOT data for lineament mapping in Hulu Lepar area, Pahang, Malaysia," European Journal of Scientific Research, vol. 34, no. 3, pp. 406–415, 2009.
38. A. Masoud and K. Koike, "Tectonic architecture through Landsat-7 ETM+/SRTM DEM-derived lineaments and relationship to the hydrogeologic setting in Siwa region, NW Egypt," Journal of African Earth Sciences, vol. 45, no. 4-5, pp. 467–477, 2006.
39. M. L. Süzen and V. Toprak, "Filtering of satellite images in geological lineament analyses: an application to a fault zone in Central Turkey," International Journal of Remote Sensing, vol. 19, no. 6, pp. 1101–1114, 1998.
40. A. Ganas, S. Pavlides, and V. Karastathis, "DEM-based morphometry of range-front escarpments in Attica, central Greece, and its relation to fault slip rates," Geomorphology, vol. 65, no. 3-4, pp. 301–319, 2005.

41. S. Sarapirome, A. Surinkum, and P. Saksutthipong, "Application of DEM data to geological interpretation: Thong Pha Phum area, Thailand," in Proceedings of the 23rd Asian Conference on Remote Sensing (ACRS ‹02), Kathmandu, Nepal, 2002.
42. G. Sarp and V. Toprak, "An Integrated Lineament Analysis from Satellite Images," in Proceedings of the 28th Asian Conference on Remote Sensing (ACRS ‹07), Kuala Lumpur, Malaysia, 2007.
43. A. Anwar, M. A. Juhari, and A. Ibrahim, "Automatic mapping of lineaments using shaded relief images derived from digital elevation model (DEMs) in the Maran—Sungi Lembing area, Malaysia," Electronic Journal of Geotechnical Engineering, vol. 15, pp. 949–957, 2010.
44. P. C. I. Geomatica, PCI Geomatica User's Guide Version 9. 1, Richmond Hill, Ontario, Canada, 2001.
45. A. Ibrahim and M. A. Juhari, Dictionary of Geological the Basic Terms, Malaysia, National University of Malaysia, Selangor, Malaysia, 1990.
46. National Water Resources Authority, Hydrogeologic Map of Taiz Area (1: 50, 000), Dar El-Yemen Hydro Consultant, 1997.
47. L. Aller, T. Bennett, J. H. Lehr, and R. J. Petty, "DRASTIC: a standard system for evaluating groundwater pollution potential using hydrogeologic settings," Tech. Rep. EPA/600/2 85/018 R.S., Kerr Enviromental Research Laboratory, Enviromental Protection Agency, Ada, Okla, USA, 1995.
48. M. J. Crozier, "Field Assessment of Slope Instability," in Slope Instability, D. Brunsden and D. Prior, Eds., pp. 103–142, John Wiley and Sons, New York, NY, USA, 1984.
49. Q. Zaruba and V. Mencl, Landslides and Their Control, Elsevier, Amsterdam, The Netherlands, 1982.
50. I. Das, S. Sahoo, C. van Westen, A. Stein, and R. Hack, "Landslide susceptibility assessment using logistic regression and its comparison with a rock mass classification system, along a road section in the northern Himalayas (India)," Geomorphology, vol. 114, no. 4, pp. 627–637, 2010.

Chapter 3

Mapping Radon-Prone Areas Using Y-radiation dose Rate and Geological Information

M García-Talavera[1], A García-Pérez[2], C Rey[1], and L Ramos[1]

[1]Consejo de Seguridad Nuclear (CSN), C/Justo Dorado, 11, E-28040 Madrid, Spain

[2]Departamento de Estadística, I. O y C. N., Universidad Nacional de Educación a Distancia, P. Senda del Rey, 9, E-28040 Madrid, Spain

ABSTRACT

Identifying radon-prone areas is key to policies on the control of this environmental carcinogen. In the current paper, we present the methodology followed to delineate radon-prone areas in Spain. It combines information from indoor radon measurements with γ-radiation and geological maps. The advantage of the proposed approach is that it lessens the requirement for a high density of measurements by making use of commonly available information. It can be applied for an initial definition of radon-prone areas in countries committed to introducing a national radon policy or to improving existing radon maps in low population regions.

INTRODUCTION

Radon ($^{222}R_n$) exposure indoors is recognised by the World Health Organization as the second most important cause of lung cancer in many countries. The main authorities in the radiological protection field have strengthened radon provisions in their principles or standards (ICRP 2009, IAEA 2011, EC 2012), driven to some extent by the epidemiological evidence made available from radon domestic exposure studies (Darby *et al* 2005, Krewski *et al* 2006, Lubin *et al* 2004). In Europe, for example, radon exposure in the home has been estimated to account for over 9% of all lung cancer deaths (Darby *et al* 2006).

From a national public health perspective, the magnitude of the radon problem largely depends on geology, specifically on the $^{226}R_a$ content and the permeability of the rock formation. Other factors that play an important role include the construction methods of buildings, and the climate.

A strategy to control radon exposure should address the whole population both at home and in the workplace, while devoting special efforts and resources to the individuals most at risk, namely those with exposures in the right tail of the lognormal distribution that typically describes radon indoor concentrations. A very high radon level can be found in a building regardless of its location, but certain areas are much more prone to give rise to high concentrations in dwellings. Consequently, the identification of such areas is a key issue in an effective regulatory control of radon exposure.

There is no consensus definition of a radon-prone area. Generically, the 2007 Recommendations of the ICRP define it as an area in which the concentration of radon in buildings is likely to be higher than the national average. In some countries, this definition is based on a given fraction of dwellings exceeding a regulatory reference level.

The most straightforward method to delineate radon-prone areas is based on actual in-house radon measurements (see, for instance, Miles

1997, Andersen *et al* 2001, Friedmann 2012).Because it requires a relatively large number of indoor radon measurements and a sampling density that is representative of the building stock, this method is only achievable nationwide when a significant fraction of all dwellings across the country has been monitored for radon.

An alternative to map radon hazard is to measure radon in soil gas. This is a good predictor of indoor radon because soil usually is by far the main entry path of radon into buildings (Nazaroff *et al* 1988). Both the Swedish and the German systems are based on the assessment of geogenic radon potential, which is derived from soil gas measurements at one metre depth and soil permeability (Åkerblom 1986, Kemski *et al* 2001).

Other indirect methods are based on the positive correlation of ^{238}U or $^{226}R_a$ concentration in surface rock or soils with radon in soil gas (Agard and Gundersen 1992), and, in turn, with radon in dwellings. Ielsch *et al* (2010) mapped the geogenic radon potential in Bourgogne (France), applying a classification scheme of geological units according to their uranium content determined by means of rock sample analysis. The suitability of equivalent uranium ($_eU$) derived from airborne surveys of γ-rays to map radon-prone areas has also been demonstrated in regions of Canada, Northern Ireland and Norway (Ford *et al* 2000, Appleton *et al* 2011, Smethurst *et al* 2008). In addition, the US EPA map of radon zones (US EPA 2013) includes aerial radiometric data on $_eU$ as one of the factors contributing to radon potential in a given area.

In Spain, although several regional surveys have identified areas of high radon exposure (Matarranz 2004), a nationwide delineation of radon-prone areas cannot be achieved by analysis of the in-house radon measurements alone because these present gaps in spatial coverage and are insufficient in number compared to the housing stock.

However, such areas need to be identified throughout the country, applying a uniform methodology, in order to prompt legislative action aimed at reducing the exposures of people residing in risk areas. As radon regulation is enacted at national level, it would be unacceptable to create regulation that is binding in some parts of the territory while setting aside other areas that are subject to similar or greater risk. In addition, it would be advantageous to delineate radon-prone areas at the smallest administrative level (which in Spain is the municipal one) since this would allow for a better use of public resources and a more efficient distribution of regulatory burdens.

In this paper, we present a methodology to map areas countrywide where 10% of residential buildings and houses exceed the reference level of 300 B_q m^{-3} set in national legislation (CSN 2011). This methodology makes use of the current radon measurement database and other available surrogate information, including the natural γ radiation map of Spain at 1:1000000 scale (Suárez *et al* 2000). We show that outdoor γ-ray exposure at 1 m above ground can contribute to successfully delineate radon-prone areas. We choose an outdoor absorbed dose rate in air of 66 ${}_{n}G_{y}$ h^{-1} as the cut-off point below which municipalities are considered non-radon-prone. Above this value, we take into account geological information and group radon measurements by lithostratigraphic unit. Each of these units is classified as radon prone or non-radon-prone based on a hypothesis test on the 90th percentile for the corresponding distribution of radon levels.

The method is proven to work well over a surface of about 500000 km^2 (mainland Spain) of very diverse geology and precipitation regime, ranging on average from less than 300 mm yr^{-1} in the southeast to more than 2000 mm yr^{-1} on the northern coast.

INPUT DATA

In order to produce the map of radon-prone areas, we merged the following data sources using ESRI ArcGis v. 10.0 (ESRI 2011):

- The national indoor radon database.
- The natural γ-radiation map.
- Geological (lithostratigraphic and metallogenic) maps.

Indoor Radon Database

The national radon database currently has over 11000 records of in-house radon measurements grouped by municipality. Several research groups have contributed to the database (Quindós *et al* 1991, Gutiérrez *et al* 1992, Amorós *et al* 1995, Baixeras *et al*1996, Pérez *et al* 1996, Baeza *et al* 2003, Quindós *et al* 2011), the bulk of measurements having been collected by the University of Cantabria through different projects sponsored by CSN. The resulting sampling density is spatially heterogeneous, tending to be higher in regions with elevated radon level and at the most densely populated areas.

Roughly 90% of measurements were made with solid-state nuclear track detectors exposed for a period from three to six months, whereas the remaining 10% were obtained with charcoal canisters exposed over three days. Although, on a house-by-house basis, short-term measurements do not provide as reliable estimates of the annual radon level as the several-month-long measurements, an overall good agreement was found in surveys conducted in all municipalities where the two types of result could be compared (García-Talavera *et al* 2013). Furthermore, because our scope was to quantify risk rather than actual population exposure, we selected measurements made at ground or first floors.

The current radon database suffers from some limitations that hamper modelling approaches: it neither reports precise measurement location (only name of the village, town or city is provided) nor contains information on relevant dwelling characteristics (such as year of construction or type of foundation). Moreover, there are a limited number of measurements in comparison to the building stock. Out of the approximately 7900 municipalities in peninsular Spain, radon measurements have been made in 1170. Yet, for roughly one-third of them, the number of available measurements (n) is 1 or 2; 452 have $7 \leq n < 26$; and 25 have $n \geq 27$.

The Natural Y-radiation Map

Many countries have developed γ-radiation maps of their territories as such maps are cost effective and have manifold applications: information on population radiation doses, documented baseline for accident and emergency situations, aid to raw materials or mineral exploration, etc.

The MARNA (Map of Natural Radiation of Spain) was undertaken as a collaborative project of CSN and ENUSA and joined later by several universities (Suárez *et al* 2000). Its main objective was to draw up a map of natural γ-radiation expressed as microroentgens per hour ($_{\mu}$R h^{-1}) at 1 m above ground.

It combined information from various sources.

- Over 150000 aerial measurements from uranium deposit exploration obtained in radiometric flights at an altitude of 120 m between 1967 and 1981. The detectors used were large volume NaI scintillators with an energy window from 0.4 to 2.8 MeV.
- Ground-based measurements, static and car-borne, also obtained in uranium exploration campaigns covering an energy window from 0.4 to 2.8 MeV.
- Over 450000 ground-based static and car-borne measurements from project-specific campaigns using Saphimo SRAT and ES3 S equipment for total counting and an Exploranium GR-1300 for spectrometric measurements.

After correcting the airborne data for cosmic radiation, a good agreement was found between airborne and ground measurements (linear correlation coefficient greater than 0.92). To draw up the map to 1:1000000 scale mean radiometric values (corresponding to a nominal cell size of 7 km×5 km) were used. The 7 km×5 km grid of regularly spaced values was submitted to a kriging interpolation using Golden Software Surfer 7. In this step, an anisotropic variogram model was specified whose parameters were locally varied to describe the orientation of geological structures. The performance of the kriging model was verified by standard cross-validation.

Geological Maps

Lithostratigraphic (LS) information was extracted from the 1:200000-scale national map developed by the Institute of Geology and Mineral Exploration (IGME 2012a). This map represents cartographic units that are delimited mainly on the basis of lithic characteristics and stratigraphic position.

The metallogenic map at 1:200000 scale (IGME 2012b) showing the nature and distribution of mineral resources was used to establish the location of uranium deposits and occurrences. The most important of the several tens of known uranium mineralised zones are located in the Schist–Greywacke Complex, a geological structure of the Central Iberian Hercynian belt. They mainly occur in fracture and breccia zones of Palaeozoic slates and granitic plutons.

STATISTICAL FRAMEWORK

We classify an area as radon prone when the 90th percentile of indoor radon concentration in its building stock is greater than the reference level. In practice, nevertheless, classification has to be done under uncertainty, as only a small sample of the dwellings in an area would typically have been measured for radon and, consequently, the point estimate of the percentile would differ from the true value of the population percentile.

There are many ways of calculating percentiles, but, given that the shape of the distribution is known, a parametric estimate would in principle be subject to less error. Because radon measurements in dwellings most often conform to a lognormal distribution, the parametric estimate is usually preferred. For a lognormal distribution, this can be obtained as a function of the sample geometric mean and the sample geometric standard deviation. If the geometric standard deviations of the areas in question do not differ statistically, the geometric means can be used instead of the percentiles for area classification.

Point estimates, however, do not provide a piece of information that hypothesis testing does: whether there are significant differences between the true value of a parameter and its estimate. Therefore, the most consistent approach to area classification is to acknowledge uncertainty by basing the decision on a statistical hypothesis test. Specifically, we have developed a hypothesis test that uses the upper bound on the 90th percentile as the critical value.

As for the rest of statistical tests included in this paper, we have used SPSS v. 19.0 (IBM Corp 2010) and the R software (R Development Core Team 2008).

The Radon-prone Hypothesis Test

If we denote by $X_{0.9}$ the 90th percentile of indoor radon concentration, X, in a given area, the problem of classifying it as radon prone or not can be formalised as a hypothesis test problem where the null hypothesis, H_0: $X_{0.9} \geq 300$ B_q m^{-3}, is tested against the alternative hypothesis, H_1: $X_{0.9} < 300$ B_q m^{-3}. The critical point for rejecting H_0 is $\tilde{X}_{0.9}$ the upper bound on $X_{0.9}$, at the α-significance level.

As it is customary, we assume that indoor radon data X, follow a lognormal distribution. We denote by m_g the sample geometric mean and by s_g the sample standard geometric deviation. Then, the random variable $Y = \log X$ is normally distributed with mean μ_y and standard deviation σ_y.

Given a random sample $\{x_i\}$, the upper bound on the 90th percentile of Y, $Y_{0.9}$, can be obtained as (Hahn and Meeker 2011)

$$\tilde{Y}_p = \bar{y} + g'_{(1-\alpha;p;n)} S_y,$$

Where $\bar{y}$ is the sample mean of $\{y_i\} = \{\log x_i\}$ $S_y = \sqrt{n/(n-1)}s_y$, n is the sample size, s_y is the standard deviation of $\{y_i\}$, and the factors $g_{(\gamma;p;n)}$ are given in table 1 of Odeh and Owen (1980).

Because of the relationship between the theoretical percentiles of X and Y, $P\{Y < Y_{0.9}\} = 0.9 \leftrightarrow P\{X < e^{Y}_{0.9}\} = 0.9$, the upper bound on the 90th percentile of X is

$$\tilde{X}_{0.9} = \exp\{\tilde{Y}_p\} = \exp\{\bar{y} + g'_{(1-\alpha;0.9n)} S_y\} = \exp\{\bar{y}\} \exp\{g'_{(1-\alpha;0.9;n)} S_y\}$$
$$= m_g s_g \exp\left\{g'_{(1-\alpha;0.9;n)} \sqrt{\frac{n}{n-1}}\right\}.$$

At the α-significance level, the null hypothesis can be rejected (thus concluding that a particular area is not radon prone) if and only if $\tilde{X}_{0.9} < 300\text{Bqm}^{-3}$. In particular, we set $\alpha = 0.10$ (probability of type-I error), which allows for a 10% probability of failing to identify a true radon-prone area.

The above equation can be applied for $n > 2$, but small sample sizes would result in large probabilities of type II error (incorrect non-rejections of H_0). To control this effect, Faulkenberry and Weeks (1968), suggested that the sample size be large enough so that

- There is a large probability (1 − α) that the tolerance bound will be exceeded by at least 100p›% of the normal population, and
- The probability (δ) that more than 100p*% of the population will exceed the said tolerance bound is small, where p› and p* are specified proportions, and p* > p›.

The probability of a successful demonstration is $p_{DEM} = Pr(p > p') \geq 1 - \delta$. For $p' = 0.90$; $p^* = 0.98$ and $\delta = 0.10$, we get a minimum sample size of 27. Consequently, we established this figure as the minimum sample size to apply the radon-prone hypothesis test.

RESULTS AND DISCUSSION

Delineating radon-prone areas at municipal level requires data on the probability of exceeding the radon reference level for every municipality in the country. Such a probability, however, cannot be based on in-house measurements alone, since these are only available at a small percentage of municipalities, with barely a few tens of them reaching the minimum sample size established to apply the radon-prone hypothesis test.

Hence it was necessary to identify whether, on a country-size scale, other environmental variables might be good surrogates for indoor radon so as to make reasonably accurate predictions at unsampled or poorly sampled locations. Specifically, we investigated outdoor γ-dose rate and geology.

Modelling Indoor Radon as a Function of Y-dose Rate

Previous research (Cascón *et al* 2002) had shown that γ-dose rate across the country exhibits a linear relationship with $^{226}R_a$ in soil, which, in turn, is known to be correlated with ^{222}Rn indoors (Nason and Cohen 1980). This fact prompted a study on the usability of the γ-radiation map of Spain to predict radon levels directly in a northwestern granitic region (Quindós *et al* 2008). Three areas with different γ-radiation levels (<65–87 $_nG_y\ h^{-1}$, 65–130 $_nG_y\ h^{-1}$, and 87–>130 $_nG_y\ h^{-1}$) were measured for radon indoors. Based on the main statistical parameters for the three corresponding radon distributions, this study concluded that γ-radiation dose rate is a qualitative indicator of high indoor radon level rather than a good quantitative predictor.

We analysed the relationship between indoor radon concentration and γ-dose rate using data from the whole Spanish mainland. In particular, for each municipality, we considered the geometric mean

(m_g) and the geometric standard deviation (s_g) of the measured radon concentrations, provided that there were more than six available measurements[3]. The corresponding average γ-dose rate was obtained from the MARNA map as follows: the built-up area of the municipality was determined using information provided by the National Centre for Geographic Information in a 1 km × 1 km grid. By a bilinear interpolation from the MARNA surfaces, the program Surfer 7 provided the γ-exposure value at the centre of every grid cell over which the built-up area spreads. We then computed the average of these values and converted it to absorbed dose in air using 1 R =0.008 76 G_y.

[3]For $n = 6$ the error associated with the geometric mean estimate is around 50% (Hewett **1995**).

Our analysis showed that there is a statistically significant but rather weak correlation between radon geometric mean (GM) and average γ-dose rate (R_γ): Pearson $r = 0.30$, $p = 0.001$; Spearman $\rho = 0.65$, $p = 0.000$.

Radon geometric standard deviation (GSD), on the other hand, is not linearly correlated either to GM ($r = 0.087$; $p = 0.082$) or to R_γ($r =- 0.060$; $p = 0.229$), although it exhibits a weak but significant rank correlation with the former variable ($\rho = 0.108$, $p = 0.030$). The distribution of GSD for all municipalities with $n > 6$ has a median value of 2.05, and a range going from 1.13 to 6.93.

The natural logarithm of GM is better correlated with R_γ ($r = 0.61$; $p = 0.000$). We performed a minimum least-squares analysis for R_γ versus ln (GM) (see figure 1), obtaining an R^2 value of 0.37. A very poor fit ($R^2 = 0.05$) was achieved when, instead of geometric means, 90th percentiles ($X_{0.9}$), assessed by the maximum likelihood parametric estimate, were considered.

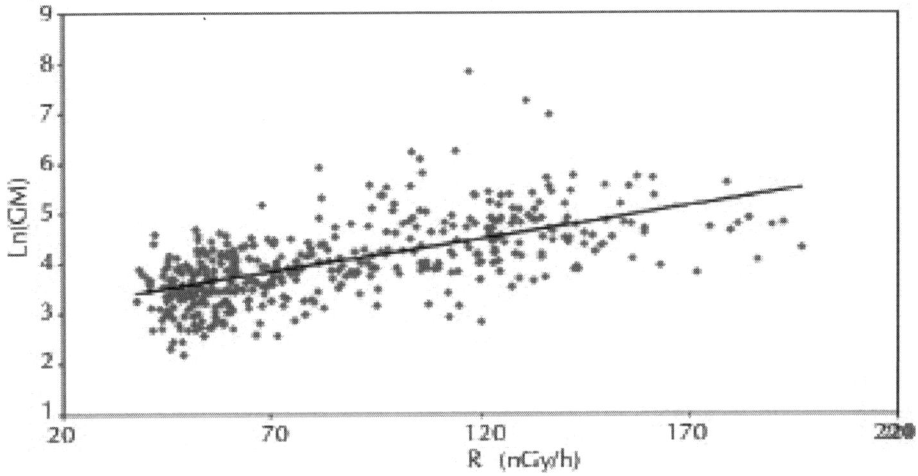

Figure 1: Linear least-squares regression fit for average γ-dose rate (R_γ) in nGy h^{-1} and natural log of geometric mean (GM) of radon concentrations in Bq m^{-3}.

None of these regression models can explain a sufficient amount of variation to allow the construction of radon maps that carry an acceptable confidence of accuracy. Yet the relevant information to label an area as radon prone is not actually m_g nor $X_{0.9}$, but the probability of them exceeding a certain threshold value. This means that, to this purpose, radon indoors can be expressed as a categorical rather than as a quantitative variable, as presented in section 4.2.

Correspondence between Categorical Variables

First, we investigated the role of γ-dose rate as a categorical variable using the same dataset as in section 4.1: we selected cut-offs of 66 $_{n}G_{y}$ h^{-1} (~7.5 $_{\mu}R$ h^{-1}) and 123 $_{n}G_{y}$ h^{-1} (~14.0 $_{\mu}R$ h^{-1}) for the γ-dose rate because the categories defined by these values provide a good division of geological units (the >7.5 $_{\mu}R$ h^{-1} surface roughly corresponds to the Pre-Mesozoic geological setting of peninsular Spain, whereas the 14.0 $_{\mu}R$ h^{-1} contour line encloses some granitic outcrops and westernmost distinct metamorphic areas of the Iberian Massif—see figure 4).

We considered the distribution of radon geometric means for each group of municipalities defined according to the γ-dose rate categories established above (see box plot in figure 2). The three of them can be described by a lognormal distribution (Kolmogorov–Smirnov test: $Z_0 = 0.514$; $p_0 = 0.954$; $Z_1 = 1.161$; $p_1 = 0.135$; $Z_2 = 1.089$; $p_2 = 0.186$). Levene's test, performed on the log-transformed data, showed heterogeneous variances ($W = 5.303$; $p = 0.005$). Thus, we applied the Kruskal–Wallis test, which revealed that the three groups are statistically different ($\chi^2 = 151.4$; $p < 0.05$).

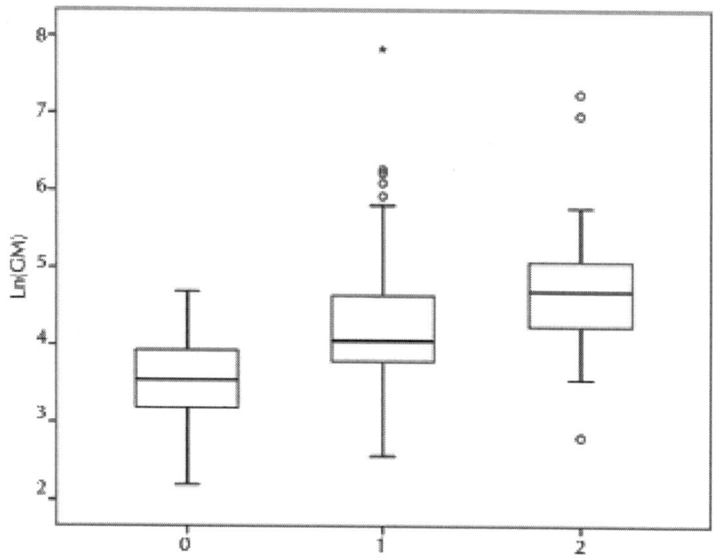

Figure 2: Box plot of natural log of radon geometric mean for municipalities subjected to the following average γ-dose rates: (0) <66 $_{n}G_{y}\ h^{-1}$; (1) 66–123 $_{n}G_{y}\ h^{-1}$; (2) >123 $_{n}G_{y}\ h^{-1}$.

Next we carried out a correspondence analysis to explore the degree of association between the following categories of γ-dose rate and indoor radon (GM).

- <66 $_{n}G_{y}\ h^{-1}$ and <70 $B_{q}\ m^{-3}$.
- 66–123 $_{n}G_{y}\ h^{-1}$ and 70–120 $B_{q}\ m^{-3}$.
- >123 $_{n}G_{y}\ h^{-1}$ and >120 $B_{q}\ m^{-3}$.

A good association was found for the three pairs (see figure 3(a)). Figure 3(b) shows a graphic representation of the associated contingency table. A geometric mean of 70 Bq m^{-3} corresponds (for n = 27 and s_g = 2.37, namely, the 75th percentile for the GSD distribution) to a 90% confidence upper bound on $X_{0.9}$ of approximately 300 Bq m^{-3}. The 120 Bq m^{-3} cut-off was selected because it achieves the closest association to the >123 nGy h^{-1} category.

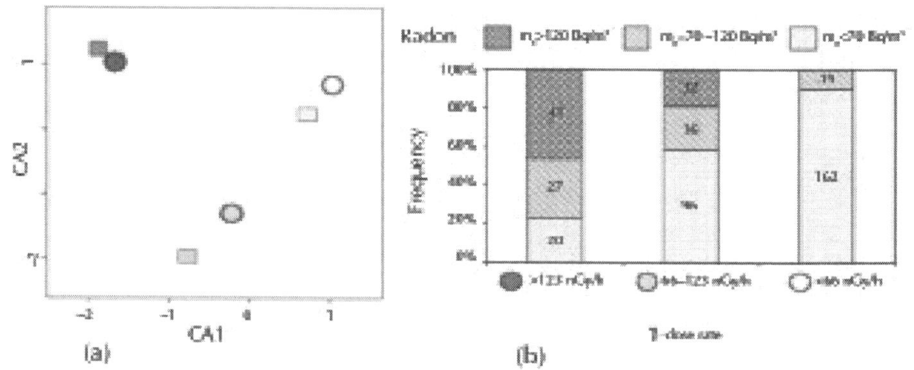

Figure 3: (a) Two-dimensional correspondence map showing a close association between the following categories of γ-dose rate and radon concentration: <66 nGy h^{-1} and <70 Bq m^{-3}; 66–123 nGy h^{-1} and 70–120 Bq m^{-3}; >123 nGy h^{-1} and >120 Bq m^{-3}. (b) Segmented bar chart representing the relationship between $R_γ$ and GM. It compares the percentage (and number of cases) that each radon category contributes to a total across categories of $R_γ$.

Out of 181 municipalities with average γ-dose rate < 66 nGy h^{-1}, nearly 90% (95% CI= (0.85, 0.94) estimated by the modified Wald method; Agresti and Coull 1998) have radon geometric means lower than 70 Bq m^{-3}. A higher proportion, 97% (95% CI= (0.93, 0.99), likewise calculated), have 90th percentiles lower than 300 Bq m^{-3}.

Therefore, selecting a cut-off point of 66 nGy h^{-1} for a screening test based on γ-dose rate can be a valid option to discriminate non-radon-prone municipalities. We considered two such screening tests using the following criteria, respectively, to define radon-prone areas: (i) m_g ≥ 70 Bq m^{-3}; (ii) $X_{0.9}$ ≥ 300 Bq m^{-3}. Table 1 show their positive predictive value (PPV, probability that a municipality is radon prone given that the test result is positive) and their negative predictive value (NPV, probability that the municipality is non-radon-prone given that the test result is negative). Both tests have a poor PPV but achieve a high NPV, notably for $X_{0.9}$.

Table 1: Positive predictive value (PPV) and negative predictive value (NPV) of a test with a cut-off point of 66 ${}_{n}G_{y}\ h^{-1}$ γ-dose rate for screening municipalities with radon $m_g \geq 70\ B_q\ m^{-3}$ and municipalities with radon $X_{90} \geq 300\ B_q\ m^{-3}$

	PPV	NPV
$m_g \geq 70\ B_q\ m^{-3}$	0.54	0.90
$X_{90} \geq 300\ B_q\ m^{-3}$	0.25	0.97

Because of the high NPV, regardless of the selected point estimate, all areas with γ-dose rate lower than 66 ${}_{n}G_{y}\ h^{-1}$ ($\sim 7.5\ {}_{\mu}R\ h^{-1}$) were directly classified as non-radon-prone see map in figure 4. Most of these correspond to Tertiary and Quaternary soils developed over calcareous sedimentary rocks.

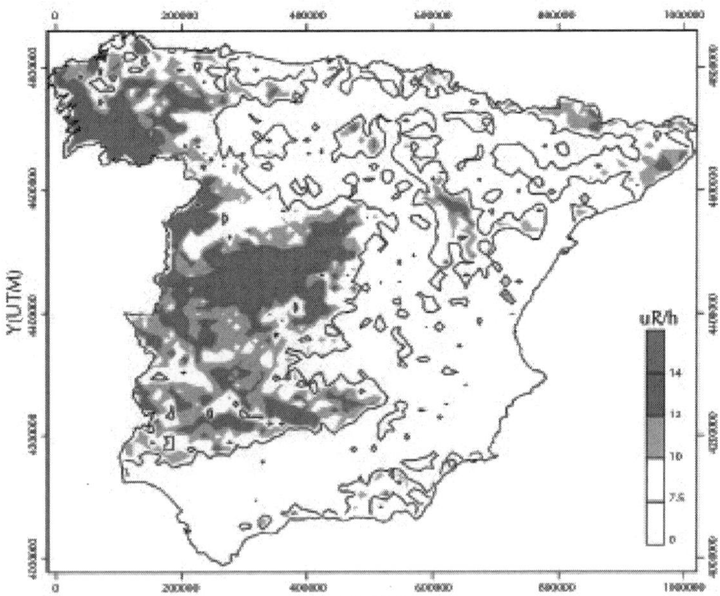

Figure 4: Map of γ-ray exposure at one metre above the ground ($R_γ$) for mainland Spain. Blank areas are directly classified as non-radon-prone.

Conversely, for areas with γ-dose rate in excess of 66 ${}_{n}G_{y}\ h^{-1}$, γ-radiation does not exhibit a strong enough control on radon indoors to make a reliable classification (see the first two bars in figure 3(b)).

Even the highest γ-dose category contains a significant proportion (~20%) of non-radon-prone municipalities.

Merging Geological Information

The control of geology on indoor radon levels is well recognised (Miles and Ball 1996, Mikšová and Barnet 2002, Kemski *et al*2002, Appleton and Miles 2010, Friedmann and Gröller 2010). Consequently, for those parts of the country with γ-dose rate in excess of 66 ${}_{n}G_{y}\ h^{-1}$ (i.e., >7.5 ${}_{\mu}R\ h^{-1}$ surface from map in figure 4), we included geological information in the analysis in order to establish radon-prone areas at geological-formation level rather than at municipality level. Specifically, we considered the lithostratigraphic (LS) map of Spain.

To analyse radon data, we selected all LS units that intersected in more than 10% of their surface with populated regions exhibiting γ-dose rate >66 ${}_{n}G_{y}\ h^{-1}$. Some of these units were further regrouped based on composition, age and permeability (see table 2). The LS unit has, on the one hand, a characteristic range of $^{226}R_a$ concentration and, on the other hand, a typical permeability[4]value, both of which have a link to indoor radon. Yet, even within the same LS type, a substantial variability of indoor radon levels could be found due to the influence of unaccounted-for variables (such as climate, dwelling characteristics, or depth to the water table).

[4]To a certain extent, permeability is determined by LS unit. However, where the predominant flow mechanism is through fractures (e.g. igneous or metamorphic rocks), local permeability can be very variable, up to several orders of magnitude, whereas more permeable lithologies, dominated by inter-granular flow, present a rather homogeneous local permeability.

Table 2: Statistical parameters of radon levels and average γ-dose rate ($\hat{R}_\gamma$) in lithostratigraphic units within regions of γ-dose rate values predominantly higher than 66 $_nG_y$ h^{-1}. Those whose upper bound on the 90th percentile ($\tilde{X}_{0.9}$) of measured radon concentrations exceeds 300 B_q m^{-3} are classified as radon prone

Unit description	**Sample size**	$\hat{R}_\gamma$ (**$_nG_y$ h^{-1}**)	m_g (**B_q m^{-3}**)	s_g	$\tilde{X}_{0.9}$ (**B_q m^{-3}**)
Hercynian acid plutonic rocks (Palaeozoic)	770	131	118.0	2.7	452.5
Hercynian arkosic sandstone and schist (Palaeozoic)	96	121	57.6	2.8	261.9
Sedimentary deposits of Tertiary age (Guadiana basin)	164	117	55.7	1.8	130.9
Slates, metagraywacke and mica-schist (Precambrian–Palaeozoic)	331	114	98.0	2.7	375.4
Metamorphic rocks (Precambrian–Cambrian–Ordovician)	85	110	131.9	2.3	459.6
Tertiary siliciclastic deposits, mainly arkosic sandstone and mudstone (Tagus–intra-Betic–Guadalquivir)	131	110	70.1	2.0	185.8
Precambrian slates and greywacke (Schist–Greywacke Complex)					
SC	119	107	105.1	2.9	493.9
SCb	94		53.1	2.0	150.6
Conglomerate/lutite/coal (Palaeozoic)	31	96	83.1	1.7	203.2
Olot volcanic rocks (Quaternary)	68	83	62.1	2.3	218.1
Palaeozoic microconglomerate, sandstone and shale (Cantabrian and Western Astur–Leonian Zones)	82	80	70.7	2.0	207.7

Alluvial basin-fill deposits of Tertiary age (Douro basin)	95	78	44.5	2.1	130.5
Quaternary basin-fill deposits along stream flood plains, salt marshes, estuaries, etc	491	75	43.6	2.5	152.7
Flysch sandstone and lutite alternations (Ebro basin and Pyrenees)	45	68	44.7	2.3	170.6
Alluvial basin-fill deposits of Tertiary age (Guadalquivir basin)	60	64	30.0	2.6	129.5

We first analysed whether radon variability within every LS unit is a random fluctuation about the mean or conforms to a spatially structured pattern. In the former case, the radon-prone hypothesis test could be directly applied on the combined raw data. The latter case would possibly require a different approach, since spatial patterns in the data could impair the performance of the standard statistical test of a hypothesis (Legendre 1993).

Because no geo-referenced information on individual radon measurements was available, we examined the variability of aggregate data expressed as the municipalities' geometric means (considering only those calculated from more than six measurements). Besides, to avoid bias in the results due to lack of precise measurement location, we excluded municipalities whose built-up area spreads over more than one LS unit. Within every LS unit, data are consistent with a lognormal distribution, as verified by the Shapiro–Wilk test at $p = 0.05$. We thus obtained the log-transformed geometric means in order to achieve normally distributed attribute data.

Unfortunately, in most LS units, the small number of municipalities, together with an irregular spatial configuration, made it impossible to conduct reliable spatial statistical analyses (it is, for instance, difficult to properly estimate a variogram with fewer than 50 data points; Olea 2006). Visual inspection did not reveal any global pattern except for slates and schist (SC) of the Schist–Greywacke Complex (see figure 5). As for the other LS units, the lack of spatial structure may be statistically confirmed on the most extensive ones (namely Hercynian acid plutonic (AP) rocks and Quaternary (Q) sedimentary material) for which sample sizes are large enough (>50 municipalities).

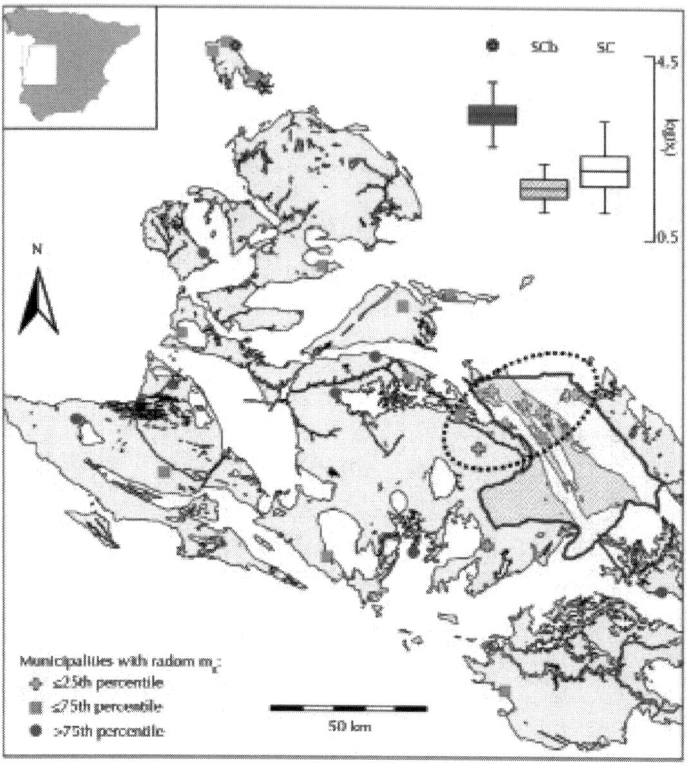

Figure 5: Regional map showing lithostratigraphic unit SC (Schist–Greywacke Complex slates and greywacke). According to the rank of their radon geometric means, sampled municipalities are displayed as follows: green crosses (<25th percentile); orange squares (25th–75th percentile); or red circles (>75th percentile). A star marks a municipality considered anomalous due to its high radon values. The ellipse with the dotted line highlights the patch of low radon values. The shaded area marks the extension of the local mountain range. The box plot represents the individual radon concentration values measured at the following types of municipality: municipality with anomalous radon data, municipalities over slates and greywacke within the shaded area (unit SC_b), and the rest of the municipalities in unit SC.

We obtained the experimental variograms for units AP and Q (figure 6). Both of them correspond to a pure nugget effect, as can be seen from the theoretical model fit. This model entails a complete lack of auto-correlation at distances larger than the lag bin size.

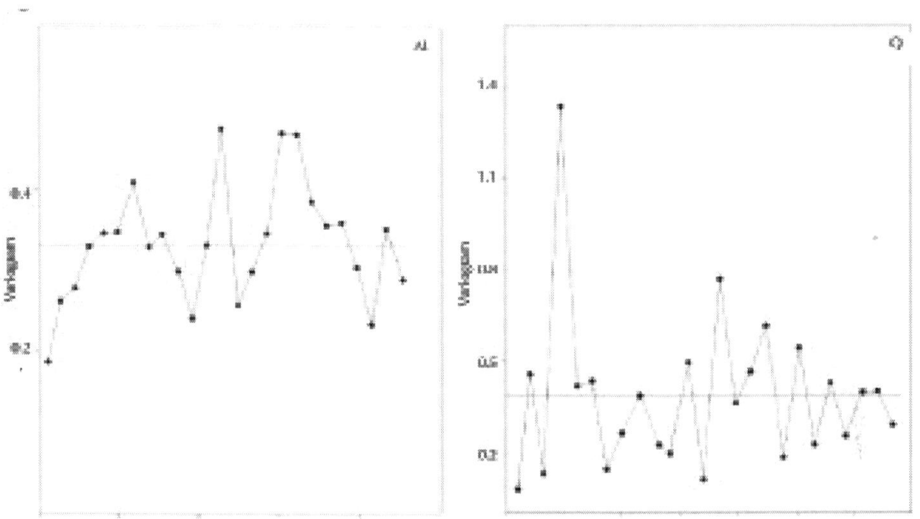

Figure 6: Experimental variograms for lithostratigraphic units AP and Q. The solid line represents the fitted pure nugget effect model.

The same results were confirmed by means of Moran's *I* (Cliff and Ord 1973). Moran's *I* ranges between −1 (perfect negative auto-correlation at global level) and 1 (perfect positive global auto-correlation), with values close to 0 indicating a random distribution. We used an inverse-distance weight matrix to compute the statistic, and a Monte Carlo permutation test (i = 1000) to assess its statistical significance. For plutonic rocks, we obtained $I_{obs} = 0.066$ ($p = 0.051$), whereas for Quaternary materials $I_{obs} = 0.039$ ($p = 0.142$). Both I_{obs} and p values indicate a lack of global auto-correlation.

The lack of a global spatial pattern (gradients or large-scale patches) suggests that precipitation and temperature regimes, which exhibit marked gradients N–S/SE through these two extensive LS units, do not exert a noticeable influence on radon indoors. Furthermore, given that the two units exhibit opposite behavior regarding radon transport (impermeable matrix with highly permeable fracture zones versus highly permeable homogeneous material), it is reasonable to assume that this result may be extrapolated to all LS units excepting unit SC, which needs further analysis.

There are two issues regarding radon data from unit SC. First, one of the sampled municipalities displays a clearly anomalous behavior (see box plot in figure 5) and was thus excluded from the study. The very high radon values (up to 15 000 B_q m^{-3}) measured in this municipality can be attributed to the existence of an underlying uranium deposit (Arribas *et al* 1983).

Although this is a unique situation, other municipalities have close uranium deposits that could potentially increase indoor radon levels. The diffusion length of this gas in soil is only about 1.2 m, but advective transport by carrier gas also occurs at long distances (several hundred metres) (Etiope and Martinelli 2002). In addition, surface soils in the proximity of uranium deposits show enhanced ^{226}Ra concentrations (up to two orders of magnitude higher than background levels), although this is a very local effect (Baeza *et al* 1994). For unit SC, we compared indoor radon data $\{x_i\}$ from five municipalities near uranium deposits (0.3–2 km) with those from the rest of the locations. The *F* test indicates equal variances for both groups ($F = 0.5821$; $p = 0.094$), whereas the two sample *t*-test proves equal means ($t = 0.5346$; $p = 0.5929$). Similar results are actually obtained when the analysis is replicated for unit AP, which also hosts several uranium deposits: $F = 0.5841$ ($p = 0.062$); $t = 0.6405$ ($p = 0.522$). Based on these results, we concluded that uranium deposits in granite and schistose rock do not have an influence on indoor radon levels at nearby population centres, at least when located at distances larger than 0.3 km (from the built-up area to the boundary of the ore deposit).

The second issue regarding radon levels in unit SC is the presence of a visually apparent large patch of minima that can be observed in figure 5. Most of the municipalities forming this patch are located in an isolated mountain range that forms a series of alternating ridgelines and valleys. They were considered as a distinct LS unit (SC_b) solely in order to detrend the data and prevent bias in statistical inference.

We then grouped the individual $\sum_{j=1}^{k_m}\sum_{i=1}^{n_j} n_{ij}$ radon measurements made in dwellings at the k_m municipalities in each LS unit m. For each of these data samples, $\{x_i\}_m$, we performed a normality test on the log-transformed values. If the sample size was larger than 50, we used the Kolmogorov–Smirnov test. Otherwise, we applied the Shapiro–Wilk test. All data sets follow lognormal distributions and, thereby, the statistical hypothesis test described in section 3.1 was performed.

Table 2 shows $\tilde{X}_{0.9}$ for the different LS units, as well as other relevant statistical parameters.

The resulting map of currently identified radon-prone areas is shown in figure 7. All units classified as radon prone have average γ-dose rates greater than 100 $_{n}G_{y}$ h^{-1}, whereas some units with average values as high as 120 $_{n}G_{y}$ h^{-1} are not radon prone. Also, at the geological unit level, γ-dose rate plays a role in determining radon risk but is not a good indicator above a certain threshold.

Figure 7: Map of radon-prone areas (shaded in pink), i.e. areas with $\tilde{X}_{0.9} < 300Bqm^{-3}$, for mainland Spain. Areas pending classification due to insufficient sample size are coloured in grey. The grey division lines mark the major water catchment areas.

FUTURE DEVELOPMENTS

Grey areas on the map (figure 7) indicate those LS_s that exhibit $R_\gamma > 66$ $_{n}G_{y}$ h^{-1} and lack enough measurements ($n < 27$) to allow calculation of the 90th percentile of radon in dwellings. With an envisaged running time of three years, the following developments are underway with the aim of completing the map shown in figure 7 and verifying its consistency with the intended application.

- Carry out a sampling campaign started in 2013 to retrieve the necessary data from all LS units in order to cover the whole territory of Spain (including the Balearic and the Canary Islands).
- Expand the study on radon spatial variability within LS units using precisely geo-reference measurements.
- Evaluate the performance of the approach presented in this paper in relation to methods based on a high density of measurements in two Spanish regions of different climates and geologies.

On the basis of such a map, CSN will launch regulation addressing the problem of populations geographically associated with high radon levels in Spain. In the long term, the radon-prone area delineation will be subject to periodic review by CSN as the radon measurement database grows, driven forward by national and local radon protection policies.

CONCLUSIONS

The method presented in this paper to delineate radon-prone areas is, from a cost-effectiveness point of view, an advantageous approach in developing a national radon policy. Because it makes use of frequently available environmental information (such as geological or γ-radiation maps), it requires a relatively small number of indoor radon measurements. This allows a more prompt implementation of protection strategies, preferentially targeted to the areas most at risk, thus optimising resources and reducing regulatory burdens.

It also shows limitations in the sense of spatial accuracy and that maps developed on actual in-house measurements would always provide the most reliable estimate. More elaborate predictive models accounting for effects such as local soil permeability or dwelling characteristics could also perform better. As conceived here, the radon-prone area map is the first version of a long-term dynamic project to be corroborated and improved as new precisely geo-referenced radon data become available from specific sampling campaigns and from radon testing required by regulation. The continuous input of radon measurements is essential to improve the accuracy of the defined areas and, consequently, to ensure better protection for individuals.

External γ-dose rate has been proved to discriminate non-radon-prone municipalities excellently. Above the cut-off of 66 ${}_{n}G_{y}$ h^{-1}, predictions on the probability of exceeding the reference level must be made not at the municipality level but at the lithostratigraphic unit level. We have proposed a hypothesis test on the 90th percentile for the radon indoors distribution that controls the probability of both false positive and false negative results. This implies acknowledging the uncertainty on making the right decision on area classification rather than relying on a point estimate.

REFERENCES

1. Agard S S and Gundersen L C S 1992 the geology and geochemistry of soils in Boyertown and Easton Pennsylvania*Field Studies of Radon in Rocks* ed L C S Gundersen and R B Wanty (Boca Raton, FL: CRC Press)
2. Agresti A and Coull B A 1998 Approximate is better than 'exact' for interval estimation of binomial proportions *Am. Stat.*52 119–26
3. Åkerblom G 1986 Investigation and mapping of radon risk areas *Geology for Environmental Planning* ed F C Wolff (Trondheim: GSN)
4. Amorós R, Bolta J M, Navarro E and Roldán C 1995 Radon levels in the Valencian community (Spain) *J. Environ.Rad.*27 125–31
5. Andersen C E, Ulbak K, Damkjær A, Kirkegaard P and Gravesen P 2001 mapping indoor radon-222 in Denmark: design and test of the statistical model used in the second nationwide survey *Sci. Total Environ.*272 231–41
6. Appleton J D and Miles J C H 2010 A statistical evaluation of the geogenic controls on indoor radon concentrations and radon risk *J. Environ. Radioact.*101 799–803
7. Appleton J D, Miles J C H and Young M 2011 Comparison of Northern Ireland radon maps based on indoor radon measurements and geology with maps derived by predictive modelling of airborne radiometric and ground permeability data *Sci. Total Environ.* 409 1572–83

8. Arribas A, Martín-Izard A and Montes J 1983 Las mineralizaciones de uranio en la zona de la Alameda de Gardón y su posición en el contexto geotectónico y metalogénico de las pizarras del oeste de la provincia de Salamanca *Studia Geol. Salmant* 18 201–24
9. Baeza A, Del Río M, Miró C and Paniagua J 1994 Natural radionuclide distribution in soils of Cáceres (Spain): dosimetry implications *J. Environ. Radioact.* 23 19–37
10. Baeza A, Navarro E, Roldán C, Ferrero J L, Juanes D, Corbacho J A and Guillén F J 2003 Indoor radon levels in buildings in the autonomous community of extremadura (Spain) *Radiat. Prot. Dosim.*103 263–8
11. Baixeras C, Font L,Robles B and Gutiérrez J 1996 Indoor radon survey in the most populated areas in Spain *Environ. Int.* 22 S671–6
12. Cascón M C, Suárez E, Quindós L, Villaseca C, García D and Fernández J A 2002 Estimación de la concentración de radón: Proyecto MARNA *ERICCA Project Meeting (Suances, 2002)*
13. Cliff A D and Ord J K 1973 *Spatial Autocorrelation* (London: Pion)
14. CSN 2011 *Guía de Seguridad 11.02 Control de la exposición a fuentes naturales de radiación* (Madrid: CSN)
15. Darby S *et al* 2005 Radon in homes and risk of lung cancer: collaborative analysis of individual data from 13 European case-control studies *Br. Med. J.* 330 223–7
16. Darby S *et al* 2006 Residential radon and lung cancer—detailed results of a collaborative analysis of individual data on 7148 persons with lung cancer and 14208 persons without lung cancer from 13 epidemiologic studies in Europe *Scand. J. Work Environ. Health* 32 1–84
17. ESRI 2011 *ArcGIS Desktop: Release 10* (Redlands, CA: Environmental Systems Research Institute)
18. Etiope G and Martinelli G 2002 Migration of carrier and trace gases in the geosphere: an overview *Phys. Earth Planet. Int.* 129 185–204
19. European Commission 2012 *Draft Euratom Basic Safety Standards Directive* Version December (Luxembourg: EC)
20. Faulkenberry G D and Weeks D L 1968 Sample size determination for tolerance limits *Technometrics* 10 343–8

21. Ford K L, Savard M, Dessau J C, Pellerin E, Charbonneau B W and Shives R B K 2000 The role of gamma-ray spectrometry in radon risk evaluation; a case history from Oka, Quebec *Geosci. Canada* 28 59–64

22. Friedmann H 2012 The Austrian Radon Project (ARP), http://homepage.univie.ac.at/harry.friedmann/Radon/onrap_e.htm (Retrieved 30/07/2012)

23. Friedmann H and Gröller J 2010 An approach to improve the Austrian Radon Potential Map by Bayesian statistics J.Environ. *Radioact.*101 804–8

24. García-Talavera M, Matarranz J L M, Gil de Mingo R, Cadierno J P and Suárez E 2013 *El mapa predictivo de exposición al radón en España Colección Informes Técnicos* (Madrid: CSN)

25. Gutiérrez J, Baixeras C, Robles B, Saez J C and Font L 1992 Indoor radon levels and dose estimation in two major Spanish cities *Radiat. Prot. Dosim.* 45 495–8

26. Hahn G J and Meeker W K 2011 *Statistical Intervals: A Guide for Practitioners* (New York: Wiley)

27. Hewett P 1995 Sample size formulae for estimating the true arithmetic or geometric mean of lognormal exposure *Am. Indust. Assoc J.* 56 219–25

28. IBM Corp 2010 *IBM SPSS Statistics for Windows, Version 19.0* (Armonk, NY: IBM)

29. Ielsch G, Cushing M E, Combes P H and Cuney M 2010 Mapping of the geogenic radon potential in France to improve radon risk management: methodology and first application to region Bourgogne *J. Environ. Rad.* 101 813–20

30. IGME 2012a *Mapa Litoestratigráfico de España a escala 1/200.000*http://mapas.igme.es/Servicios/default.aspx#IGME_Litoestratigrafico200 (Retrieved 20 Jan 2012)

31. IGME 2012b *Mapa Metalogenético de España a escala 1:200.000 (Serie antigua)*www.igme.es/internet/cartografia/cartografia/metalogeneticoA200.asp (Retrieved 20 Jan 2012)

32. International Commission on Radiological Protection 2009 Statement on Radon Approved by the Commission in November *ICRP* Ref 00/902/09

33. International Atomic Energy Agency 2011 *Radiation Protection and Safety of Radiation Sources: International Basic Safety Standards (IAEA Safety Standards Series No. GSR Part 3 (Interim))* (Vienna: IAEA)
34. Kemski J, Klingel R, Siehl A and Stegemann R 2002 Radon transfers from ground to houses and prediction of indoor radon in Germany based on geological information. 7th International Symposium on the Natural Radiation Environment (NRE-VII) *Radioact Environ.*7 820–32
35. Kemski J, Siehl A, Stegemann R and Valdivia-Manchego M 2001 Mapping the geogenic radon potential in Germany *Sci. Total Environ.*272 217–30
36. Krewski D *et al* 2006 A combined analysis of North American case-control studies of residential radon and lung cancer *J. Toxicol. Environ. Health* A 69 533–97
37. Legendre P 1993 Spatial autocorrelation: trouble or new paradigm? *Ecology* 74 1659–73
38. Lubin J H *et al* 2004 Risk of lung cancer and residential radon in China: pooled results of two studies *Int. J. Cancer* 109132–7
39. Matarranz J L 2004 *Concentraciones de radón en viviendas españolas Colección Informes Técnicos* (Madrid: CSN)
40. Miles J 1997 Development of maps of radon-prone areas using radon measurements in houses *J. Hazard Mater.*61 53–8
41. Miles J and Ball K 1996 Mapping radon-prone areas using house radon data and geological boundaries *Environ. Int.* 22S779–82
42. Mikšová J and Barnet I 2002 Geological support to the national radon programme (Czech Republic) *Bull Geosci.* 77 13–22
43. Nason R and Cohen B L 1980 Correlation between ^{226}Ra in soil, ^{222}Rn in soil gas, and ^{222}Rn inside adjacent houses*Health Phys.* 52 73–4
44. Nazaroff W W, Moed B A and Sextro R G 1988 Soil as a source of indoor radon: generation, migration, and entry *Radon and its Decay Products in Indoor Air* ed W W Nazaroff and A V Nero (New York: Wiley-Interscience)
45. Odeh R E and Owen D B 1980 *Tables for Normal Tolerance Limits, Sampling Plans and Screening* (New York: Marcel Dekker)

46. Olea R 2006 a six-step practical approach to semivariogram modelling *Stochast. Environ. Reas. Risk Assessm.*20 307–18

47. Pérez J M, Álvarez M C, Dopico M T and Garzón L 1996 Indoor ^{222}Rn concentrations in central Asturias *Health Phys.* 70689–94

48. Quindós J L, Fernández P L and Soto J 1991 National survey of indoor radon in Spain *Environ.Int.*17 449–53

49. Quindós L S, Fernández P L, Sainz C, Fuente I, Nicolás J, Quindós L and Arteche J 2008 Indoor radon in a Spanish region with different gamma exposure levels *J. Environ. Radioact.* 99 1544–7

50. Quindós L, Baixeras C and Barros J 2011 *Ampliación del mapa español de radón.Final report to CSN* (Madrid: CSN)

51. R Development Core Team 2008 *R: A Language and Environment for Statistical Computing* (Vienna: R Foundation for Statistical Computing)

52. Smethurst M A, Strand T, Sundal A V and Rudjord A L 2008 Large-scale radon hazard evaluation in the Oslofjord region of Norway utilizing indoor radon concentrations, airborne gamma ray spectrometry and geological mapping *Sci. Total Environ.*407 379–93

53. Suárez E, Fernández J A, Baeza A, Moro M A, García D, Pozo J M and Lanaja J M 2000 Proyecto MARNA. Mapa de radiación gamma natural *Colección Informes Técnicos* (Madrid: CSN)

54. US EPA 2013 EPA map of radon zones www.epa.gov/radon/zonemap.html (Retrieved 09 Jan 2013)

Chapter 4

Geophysical Assessments of Renewable Gas Energy Compressed in Geologic Pore Storage Reservoirs

Said Attia al Hagrey, Daniel Köhn, and Wolfgang Rabbel

Department of Geophysics, Institute of Geosciences, University of Kiel, 24118 Kiel, Germany

ABSTRACT

Renewable energy resources can indisputably minimize the threat of global warming and climate change. However, they are intermittent and need buffer storage to bridge the time-gap between production (off peak) and demand peaks. Based on geologic and geochemical reasons, the North German Basin has a very large capacity for compressed air/gas energy storage CAES in porous saltwater aquifers and salt cavities. Replacing pore reservoir brine with CAES causes changes in physical properties (elastic moduli, density and electrical properties) and justify applications of integrative geophysical methods for monitoring this energy storage. Here we apply techniques of the elastic full waveform inversion FWI, electric resistivity tomography ERT and gravity to map and quantify a gradually saturated gas plume injected in a thin deep saline aquifer within the North German Basin.

For this subsurface model scenario we generated different synthetic data sets without and with adding random noise in order to robust the applied techniques for the real field applications. Datasets are inverted by posing different constraints on the initial model. Results reveal principally the capability of the applied integrative geophysical approach to resolve the CAES targets (plume, host reservoir, and cap rock). Constrained inversion models of elastic FWI and ERT are even able to recover well the gradual gas desaturation with depth. The spatial parameters accurately recovered from each technique are applied in the adequate petrophysical equations to yield precise quantifications of gas saturations. Resulting models of gas saturations independently determined from elastic FWI and ERT techniques are in accordance with each other and with the input (true) saturation model. Moreover, the gravity technique show high sensitivity to the mass deficit resulting from the gas storage and can resolve saturations and temporal saturation changes down to ±3% after reducing any shallow fluctuation such as that of groundwater table.

INTRODUCTION

One unprecedented challenge facing the human being is the energy resources, and its coupling with global climate changes and warming from greenhouse gases (GHG). Mitigation of anthropogenic GHG, including CO_2 emissions in the terrestrial atmosphere demands developments of viable alternative of renewable energy resources including hydroelectric, biomass, solar, wind, marine (wave/tides) and geothermal sources. Most of these sources produce energy only when suitable weather conditions are prevailing and not when energy directly demanded. These sources are intermittent and need buffer storage to bridge the time-gap (disparity) between off-peak production and demand peaks. The underground geology offers an adequate option for short- and long-term energy storage such as compressed air or gas energy storage, CAES (e.g., Crotogino et al. 2001; Succar and Williams 2008). The North German Basin delivers favourable conditions (geological, geochemical) for underground space utilization and has a huge capacity for CAES in porous brine aquifers and salt caverns (natural and artificial). Advantages of renewable energy storage are (1) balancing power demand and fluctuating renewable energy production, (2) bridging temporal mismatch between renewable energy production (off-peaks) and demand (peaks), i.e., storing off-peak energy supply to use it during peak demand periods, and (3) offering large buffer capacity to meet any disruptions in energy supply.

Figure 1 shows how a renewable electricity system could supply actual electricity demand during one week in Minnesota (Makhijani et al. 2012). The daily electricity demand is constantly changing while the surplus renewable generation is put into storage. The basic approach of CAES is as follows: When electricity generation is greater than demand, the energy surplus is used to compress air in the geostorage. When generation is less than demand, pressurized air is withdrawn from storage and used to drive an electricity turbine.

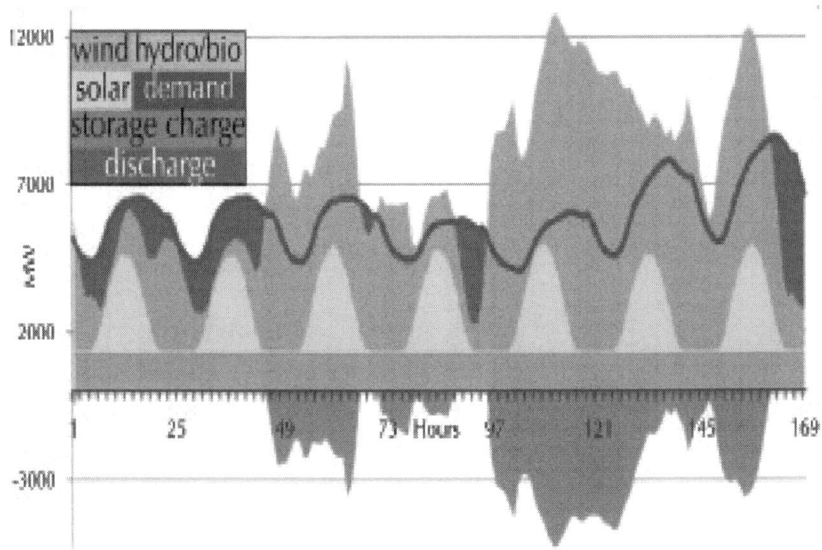

Figure 1: Daily supply and demand with storage of renewable energy, 11–17 July 2007, Minnesota (Makhijani et al.2012).

In a geological gas storage in saline formations, the gas replaces the pore brine causing strong changes in elastic moduli, density and electric resistivity. These physical contrasts justify the application of integrative geophysical techniques for monitoring this geostorage. These include techniques of elastic full waveform inversion (FWI), electric resistivity tomography (ERT), gravity and electromagnetic induction (EMI) of time- (TEM) and frequency-domain (FEM). EMI techniques (ground and air based) are applied usually for monitoring shallow targets, e.g., leakages in groundwater.

Since some years ago Germany practices a turnaround in the energy policy (German "Energiewende") and is currently leading in the production of solar and wind energy (IEA 2013). Wind energy are produced mainly at the coastal areas (on-shore and off-shore) of North Germany which is characterized by high wind speeds. We started 2012 an interdisciplinary joint research project ANGUS+ dealing with impacts of using geologic subsurface as a thermal, electric or material storage in context with alternative energy resources (Bauer et al. 2013). This includes dimensioning, risk analyses and impact predictions as a base for future space planning of the subsurface. Our main task is to develop a geophysical monitoring strategy using integrative approach of geophysical techniques (FWI, ERT, EMI and gravity) on almost realistic scenarios in the North German Basin.

We show here results of numerical simulations of elastic FWI, ERT and gravity techniques in mapping CAES reservoirs with a continuous gradual desaturation with depth. These simulations are applied for synthetic data (before and after adding 3% random noise error) and inverted using constrains on the initial models.

Gas Storages in the North German Basin

For a more realistic modelling scenario we selected a synthetic site in the North German Basin for this numerical study of CAES in geologic formations. The model block of this site ($29*28*5.5$ km^3) consists of a thick succession of 14 sedimentary layers ranging in age from Permian to Tertiary (Figure 2, Baldschuhn et al. 2001; Hese 2012). The succession shows nearly horizontal layering with a gentle anticline fold in its southern part. It shows an unconformity, where formations of the Lower cretaceous, Lias and Rhaet disappear within the anticline crest at the depth interval of 0.7–1.0 km. The succession includes two thin brine reservoirs of a porous sandstone (5–30 m thickness), namely the Rhaet (1–1.5 km depth) and Quickborn (2–2.5 km depth) formations. Both are potential pore reservoirs for CAES and only the shallow Rhaet formation is considered in this study. The compressed gas is injected in the uppermost formation part within the northern anticline limb. Here this very light gas (e.g., air density of 1.29 kg/m^3 at STP "standard temperature, 0°C, and pressure, 101325 Pa") replaces the dense brine of >1100 kg/m^3 (with total dissolved solids, TDS, of >100 g/l at this depth) in the pores of the sandstone reservoir. Accordingly, the compressed gas saturation approaches its maximum value directly below the injection level and decreases gradually with depth within the reservoir of the dipping anticline flank. This downward gradual gas desaturation may correspond well with the almost realistic CO_2 plume scenario injected in a deep saline aquifer (Graupner et al. 2011). A combined simulation of multiphase flow, transport and geochemical reactions in a brine reservoir shows that the gas phase saturation decreases with increasing distance away from the injection point. The limited lateral extension of the thin Rhaet reservoir leads to dominating the downward gas propagation, where the porosity and permeability decrease generally with depth. For this reservoir we applied two saturation distributions (Sg_1 and Sg_2) gradually decreasing downwards of the range 0.60–0.27 and 0.80–0.36, respectively (Figure 2c). The applied gradual desaturation with depth (z) follows this function: $S_g = S_{g0} e^{-5(z-z_0)^2}$, with S_{g0} = maximum S_g at $z_0 = 1$ km, 1 km $> z \leq 1.4$ km. Both of S_{g1} and S_{g2} approach their maximum values near to the anticline crest (at 1 km depth) and gradually decrease with depth according to this function approaching their minimum values at 1.4 km depth.

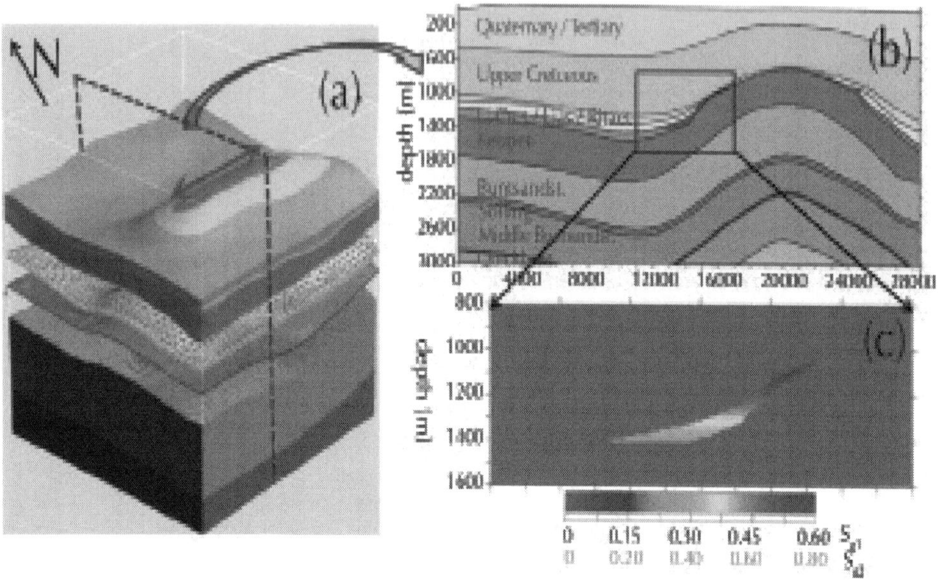

Figure 2: Study site within the North German Basin. (a) 3D block below the Quaternary-Tertiary overburden, (b) 2D stratigraphy section perpendicular to the main anticline structure, and (c) 2D saturations of compressed gas (S_{g1} and S_{g1}) injected in the saline Rhaet reservoir of the northern limb, (partly from Hese 2012).

The Seismic Forward Problem, FWI and Model Parameterization

The propagation of seismic waves in an isotropic elastic medium can be described by a system of coupled first order partial differential equations (Landau and Lifschitz 1986)

$$d\frac{\partial v_i}{\partial t} = \frac{\partial \sigma_{ij}}{\partial x_j} + f_i, \tag{1}$$

$$\frac{\partial \sigma_{ij}}{\partial t} = \lambda\left(\frac{\partial \epsilon_{11}}{\partial t} + \frac{\partial \epsilon_{22}}{\partial t} + \frac{\partial \epsilon_{33}}{\partial t}\right)\delta_{ij} + 2\mu\frac{\partial \epsilon_{ij}}{\partial t} + \frac{\partial T_{ij}}{\partial t}, \tag{2}$$

$$\frac{\partial \epsilon_{ij}}{\partial t} = \frac{1}{2}\left(\frac{\partial v_i}{\partial x_j} + \frac{\partial v_j}{\partial x_i}\right), \tag{3}$$

where d denotes the density, v_i the particle velocity, σ_{ij} the stress tensor, ϵ_{ij} the strain tensor, λ and μ the Lamé parameters, δ_{ij} the Kronecker Delta, f_i, T_{ij} the source terms for body and surface forces, respectively.

Equation (1) is a general expression for the conservation of momentum in a continuum. It is independent of the medium state - such as gas, fluid or solid. To describe the behavior of the material correctly a relationship between the forces (stresses σ_{ij}) acting on the medium and the resulting deformation (strain ϵ_{ij}) is required. For small forces/deformations and an isotropic medium this relationship is linear (generalized Hooke's law) depending only on the distibution of two material parameters λ and μ (Lamé parameters). Assuming that these parameters are time-independent the stress–strain relationship can be replaced by the stress–strain-rate equation (2). For a given isotropic elastic medium equations (1–3) can be solved numerically and therefore synthetic seismograms for any acquisition geometry calculated. Based on the solution of the seismic forward problem a high-resolution imaging concept called full waveform inversion (FWI) has been developed in the 1980s by Tarantola (1986). Since then the FWI is significantly improved and applied to a wide range of field applications (Virieux and Operto 2009).

The aim of (spatial) FWI is to minimize the data residuals $\delta u = u^{mod} - u^{obs}$ between the modelled data u^{mod} and the field data u^{obs} to deduce high resolution models of the elastic material parameters in the underground. To solve this nonlinear optimization problem an appropriate objective function E has to be defined. Similar to Asnaashari et al. (2013a), we use the following objective function

$$E = \frac{1}{2}\left(||\boldsymbol{\delta u}||_2 + \lambda_1||\boldsymbol{W}_{\boldsymbol{m}}\left(\boldsymbol{m}-\boldsymbol{m}_{\boldsymbol{prior}}\right)||_2\right) \tag{4}$$

Where the term $\frac{1}{2}\|\delta u\|_2$ denotes the L2-norm of the data misfit and $\frac{\lambda_1}{2}\left\|W_m\left(m - m_{prior}\right)\right\|_2$ a weighted L2-norm of the difference between the model parameters m and prior model information mprior used for model regularization. The parameter λ1 balances the contributions of the data misfit and the model regularization term, while the spatial variable weighting factor Wm defines which parts of the model are updated during the inversion process. The spatial weighting of the model updates is crucial for a successful inversion, because near surface inversion artefacts can introduce artificial data residuals not present in the real time-lapse data and therefore lead to an increase of the nonlinearity of the inverse problem. Like Asnaashari et al. (2013b) the magnitudes of the spatial weighting factors are based on elastic reverse-time migration (RTM) results to restrict model updates to the storage formation. The objective function equation (4) can be minimized by iteratively updating the model parameters mn (P-wave velocity (Vp), S-wave velocity (Vs), density (d)) at iteration step n, starting with an initial background model m0 using the Newton method (Nocedal and Wright 2006):

$$\boldsymbol{m}_{n+1} = \boldsymbol{m}_n - \tau_n \left(\boldsymbol{H}_{\boldsymbol{m}}^{-1} \boldsymbol{G}_{\boldsymbol{m}}\right)_{\boldsymbol{n}} \tag{5}$$

with the gradient of the objective function equation (4) with respect to the elastic model parameters

$$\boldsymbol{G}_m = \frac{1}{2}\left(\frac{\partial||\boldsymbol{\delta u}||_2}{\partial \boldsymbol{m}}\right) + \boldsymbol{\lambda}_1 \boldsymbol{W}_{\boldsymbol{m}}{}^2\left(\boldsymbol{m} - \boldsymbol{m}_{\boldsymbol{prior}}\right) \tag{6}$$

and H_m the second derivative of the objective function (Hessian). An explicit calculation of the Hessian in the time-domain is computational very expensive. Therefore, we use the quasi-Newton L-BFGS (Limited-memory Broyden-Fletcher-Goldfarb-Shanno) technique (Nocedal and Wright 2006; Brossier 2011), where the product of the inverse Hessian $H_m{}^{-1}$ with the gradient G_m is iteratively approximated by finite-differences.

The effective calculation of the time-domain gradient directions $\left(\frac{\partial\|\delta u\|_2}{\partial}\right)$ with the adjoint method for different model parameterizations are described in Tarantola (1986), Mora (1987), Shipp and Singh (2002) and Köhn et al. (2012). The step length τ_n is estimated by a line-search satisfying the Wolfe-conditions (Nocedal and Wright 2006) to assure a fast and accurate convergence of the L-BFGS algorithm. While Asnaashari et al. (2013a) introduce another regularization term to assure model smoothness, we apply a weak wavenumber domain filter to the estimated search directions at every iteration step for the same purpose.

In elastic time-lapse FWI, the data residuals are modified according to

$$\delta\mathbf{u} = \left(\left\{u^{mod}(t_1) - u^{mod}(t_0)\right\} - \left\{u^{obs}(t_1) - u^{obs}(t_0)\right\}\right), \tag{7}$$

which denote the difference between the modelled and the field data at time steps t_0 (baseline model) and t_1 (Denli and Huang 2009). This redefinition of the data residuals leads to a much stronger focusing of the model updates at reservoir level.

Based on the distribution of the air within the storage formation with a maximum gas saturation of 80% (Figure 2), an elastic model of the underground before and after the CAES injection is built. The elastic properties of the rock matrix (P-wave velocity $V_{p,m}$, S-wave velocity $V_{s,m}$, density d_m and porosity Φ) are linked with the physical parameters of the fluid and gas phases based on realistic matrix parameters to derive effective medium parameters.

The effective seismic velocities V_p and V_s and the bulk density d_b of the 100% brine saturated aquifer before the CAES injection can be described by the following averaging equations (Gassmann 1951):

$$d_b = (1-\Phi)d_m + \Phi d_{br} \tag{8}$$

$$\mu_b = \mu_d = (1-\Phi)\,\mu_m \tag{9}$$

$$K_b = K_d + (1-(K_d/K_m))^2/(\Phi/K_w + (1-\Phi)/K_m - K_d/K_m^2) \tag{10}$$

$$K_d = (1-\Phi)\,K_m \tag{11}$$

$$K_w = K_{br} \tag{12}$$

$$V_p = sqrt(K_b/d_b) \tag{13}$$

$$V_s = sqrt(\mu_b/d_b) \tag{14}$$

Where d denotes the density, *K* the bulk moduli and μ the shear moduli. The subscripts *b,m,d,br,g*and *w* mean the bulk, rock matrix, dry rock, brine phase, gas phase and wetting phase, respectively. For the multiphase flow case after the CAES injection, equations (8) and (12) are adapted to:

$$= (1-\Phi)\, d_m + \Phi\, \left(\left(1-S_g\right)d_{br} + S_g\right. \tag{15}$$

$$K_w = S_g K_g + \left(1-S_g\right) K_{br} \tag{16}$$

Based on this rock model Figure 3 shows changes of the material parameters due to the gas injection. Overall the changes of the effective medium parameters within the storage formation are quite small with maximum value changes of -240 m/s for V_p, +86 m/s for V_s and -168 kg/m^3 for *db*. The small variations of the S-wave velocity are due to density variations only, while the shear modulus remains constant.

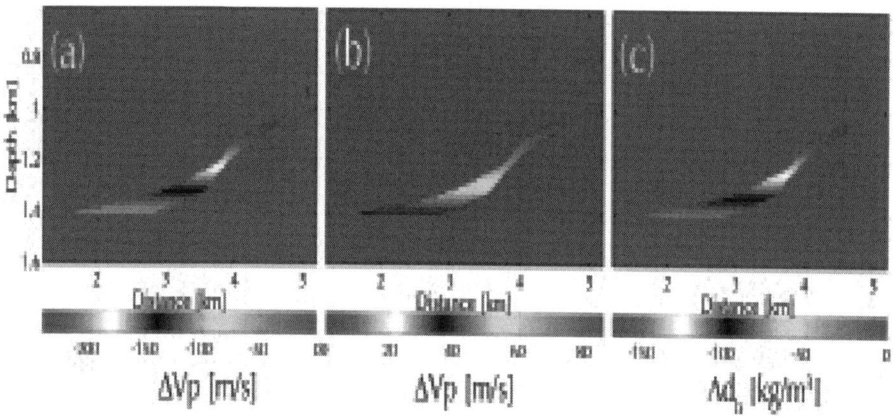

Figure 3: Input (true) changes in elastic parameters due to the gas injection in the Rhaet reservoir below the study site as obtained from the rock model. (a)ΔV_p, (b)ΔV_s and (c) density (Δd_b).

Synthetic data of a reflection seismic survey along the transect is computed by solving the 2D isotropic elastic equations of motion equations (1–3) with a time-domain finite-difference (FD) technique on a Cartesian grid (Virieux 1986; Holberg 1987; Levander 1988). The reflection seismic acquisition geometry consists of 500 vertical component geophones located at the surface. For the synthetic dataset 100 shots using a vertical impact source are recorded. The source signature is a 20 Hz Ricker wavelet. Model dimensions along the transect are 6 km length and 1.8 km depth. Using an 8th order spatial

FD operator, the model can be discretized with 600 × 180 grid points with a spatial grid point distance of 10 m. The time is discretized using $\Delta t = 1$ ms, thus for a recording time of $t = 4$ s, approximately 4000 time steps are computed. A free surface boundary condition is assumed on top of the model, while Convolutional Perfectly Matched Layers, CPMLs (Komatitsch and Martin 2007) are used at all other boundaries. The synthetic seismic sections are the input data for the elastic FWI. Figure 4 shows the common shot-gather for shot 50 of the baseline model and the time-lapse data between t_1 and t_0 amplified by a factor 10. Notice the strong spurious multiple reflections due to the free boundary condition present in the baseline- and time-lapse data.

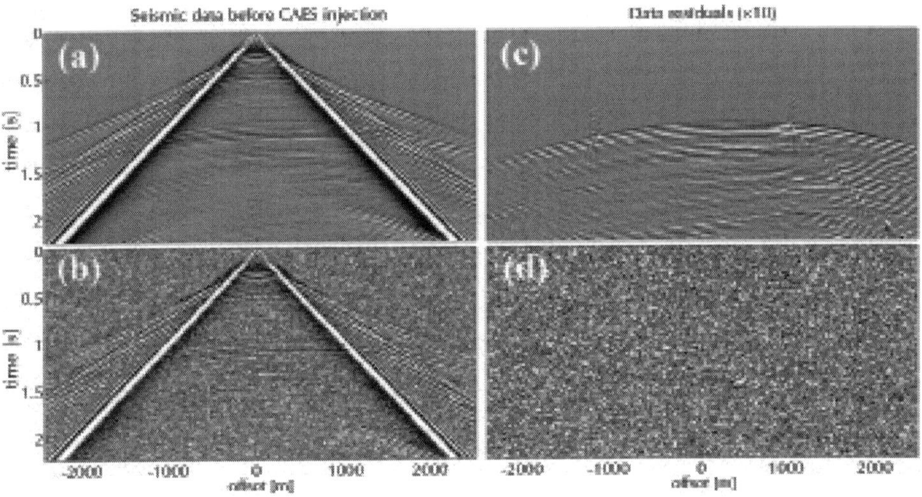

Figure 4: Common shot-gather for shot 50. (a, b) Baseline model and(c, d) time-lapse data residual between t_1 and t_0 amplified by a factor 10.(a, c) Noise-free data and (b, d) noisy data with S/N = 100.

To test the robustness of the elastic FWI approach for real field data applications we also investigate the influence of noise and added Gaussian noise within the frequency range of the source signal (0–40 Hz) with a signal-to-noise ratio S/N = 100 (1% noise) using the SUADDNOISE program of Seismic Unix (Cohen and Stockwell 2008). The resulting shot gathers are shown in Figure 4b,d for the baseline and time-lapse data, respectively.

ERT Modelling and Parameterization

At first we introduce briefly the approach for optimized electrode arrays in boreholes applied here. Like surface surveys, ERT data acquisition between two borehole electrode arrays can be conducted in the tripotential quadrupole configurations α (CPPC, C = current electrode, P = potential electrode), β (CCPP) and γ (CPCP). For an N collinear multi-electrode array, a whole comprehensive data set consists of $[N(N-1)(N-2)(N-3)/8]$ independent non-reciprocal quadrupole configurations (Noel and Xu 1991). The effective comprehensive data set results from excluding the redundant configurations of less stable inversions from the whole set, i.e., γ configurations and those of very large geometric factors (Loke et al. 2010). Resulting comprehensive data set is still huge, e.g., a pair of 32 borehole electrodes yield $>10^6$ data points. It can map subsurface targets with the highest possible resolution but at very long acquisition times (i.e., poor temporal resolution) and at high costs. Therefore, an optimization approach is based on the model resolution matrix and searches for electrode configurations that maximize the resolution of survey results (e.g., Stummer et al. 2004; Wilkinson et al. 2006). An optimized borehole data sets of practical sizes (15,000 data points) of only 1.5% of the comprehensive data set but with almost the same spatial resolution is generated in this study (e.g., al Hagrey 2012a). Comparative applications of diverse configurations (standard and non-standard) show the superiority of the optimized array results (al Hagrey 2012b). Moreover, assessing our optimized array by the technique of region of investigation index (ROI) showed that its inverted tomograms are best constrained by the data coverage in comparison to that of the other configurations (Oldenburg and Li 1999). All these confirm the effectiveness of our optimization approach applied here to generate a practical optimized data set of high resolution.

The 1storage Rhaet formation consists of a highly resistive matrix (e.g., sandstone) and conductive pore brine saturant. The bulk resistivity (ρ) resulting from the gas displacing the brine is predicted using Archie's law (Archie 1942):

$$\rho = \frac{a\rho_{br}}{\Phi^m \left(1-S_g\right)^n} \tag{17}$$

where ρ_{br} is the brine resistivity, Φ the porosity, S_g the gas saturation and *a,m* and *n* are Archie constants. The separate phases (matrix, brine and gas) are assumed without any interaction.

The electrical conductivity of the storage formation is caused mainly by the electrolytes of its pore brine. In the North German Basin, temperature and pressure, and particularly the salinity or total dissolved solids (TDS) increase with depth. Increasing both TDS and temperature causes a dramatic decrease in the resistivity (e.g., Arps 1953; Schlumberger 1985). The TDS rise increases the number of ions carrying electrical currents. The temperature rise increases the salt solubility and decreases the brine viscosity which in turn enhances the ion mobility. The pressure increase with depth, on the other hand, causes a slight increase in the resistivity due to the closure of cracks that are often filled with conductive fluids. However, this effect decreases with increasing depth and is negligible at pressure >0.3 GPa (e.g., Brace et al. 1965). In formations of the North German Basin, the average vertical gradient (with depth) of brine salinities, temperature and pressure approach 100 mg/L/m, 0.03°C/m and 22.6 kPa/m, respectively (e.g., Magri et al. 2009).

The target layers host borehole electrode arrays at 620 m offsets within a depth range of 0.9–1.8 km. Each array consists of 32 electrodes at 20 m spacing. The electric resistivity of the 2D models are parameterized by the bulk rock resistivity values calculated from Archie equation (17). Here we applied the values of 0.2 for Φ, 0.08 Ωm for ρ_{br} (corresponding to $TDS \approx 100$ g/l at 1 km depth of the North German Basin) and 1, 2 and 2 for constants *a,m* and *n* of, respectively, as typical values for the sandstone aquifer. Values of gas saturation are calculated by a potential function simulating their gradual damping with depth.

A 2.5D forward and inverse ERT modelling is carried out using modern codes (RES2DMOD, RES3DMOD × 64 and RES2DINV × 64) based on algorithms by e.g., Loke et al. (2003). The forward modelling code is applied to generate synthetic data sets between each adjacent pair of borehole electrode arrays (including inhole and crosshole) using optimized electrode configurations. These synthetic data sets are generated after gas injection in the brine reservoir of Rhaet formation. The data quality (0.6% average simulation error) is confirmed by results of tests on a homogeneous model with a constant ρ value. The technique robustness in the field is realized by adding a random error of commonly 3% to data sets in addition to their forward simulation error of 0.6%.

In the ERT inversions, diverse setup constraints (mainly regularizations) are applied. These include the minimization methods of least squares (L_2) or robust blocky normalization (L_1), and initial models of a constant homogeneous resistivity or an approximate inverse model (e.g., Claerbout and Muir1973). Two synthetic data sets (generated before and after adding 3% random noise) are inverted with incorporating mapping data of subsurface stratigraphy from prior (seismic) surveys (see next sections). Each of these two data sets was inverted twice, once by incorporating layer interfaces and once by fixing resistivity regions, both are outside the reservoir layer.

Gravity Modelling and Parameterization

Rock densities depend on the mineral composition, porosity and its content (fluid and gas), pressure$p$, temperature (T), deformations etc. The stratigraphy and average densities applied in the 3D gravity modelling of the study site are based on borehole measurements and data base of Geotectonic Atlas of northwestern Germany (Inselmann 1985; Baldschuhn et al. 2001; Hese 2012). The injected compressed gas in saline reservoirs displaces pore brine and causes a drop of the bulk density which in turn causes a decrease in the gravity components and gradients. The bulk reservoir density (d_b) of partial gas saturations is given by equation (15), whereas dry air density is given by:

$$d_g = p/(RT) \tag{18}$$

where R is the gas constant = 287.058 J/(kgK) for dry air. For calculating air and bulk densities (equations 18 and 15), we considered almost realistic values for p (average 27.786 GPa) and T(average 44°C) prevailing at 1–1.4 km depth range within the Rhaet reservoir of the North German Basin. This T value results from average local T at sea level (8°C) plus T fraction caused by the geothermal gradient down to 1.2 km depth (36°C). For instance, the average brine and air densities within this depth range approach 1123 and 303 kg/m^3, respectively, and the study sandstone reservoir will suffer from a bulk density drop up to -126 and -168 kg/m^3 corresponding to air saturations of Sg_1 for Sg_2, respectively. Figure 5 shows the 3D distributions of the gas saturation (S_{g1} for S_{g2}) and bulk density (resulting from the parameterization) in the brine reservoir of the anticline flank within the North German Basin.

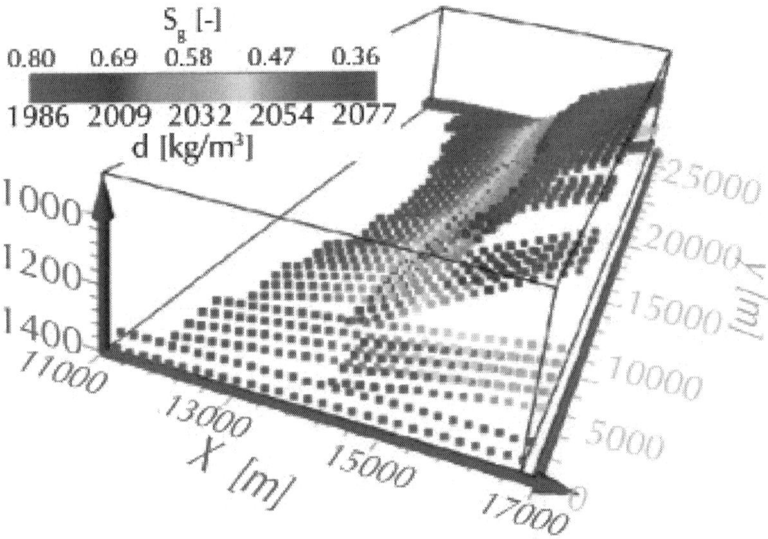

Figure 5: 3D distribution of gas saturation (S_g) and bulk density (*d*) in the brine reservoir of the anticline flank (see Figure2c).Reference *d* of reservoir (i.e., at $S_g = 0$) is 2154 kg/m³.

We used here the software IGMAS+ (Interactive Gravity and Magnetic Application System) designed for 3D gravity, gravity gradient and magnetic modelling (e.g., Götze and Lahmeyer 1988; Schmidt et al. 2011). The model is extrapolated outside the volume of interest in all directions (about two times the model length) to avoid any edge effects. We calculated the gravity field components (g_z, g_y and g_z) and gradients (g_{zx}, g_{zy} and g_{zz}) before and after CAES injection, respectively, as well as their difference (residual) anomalies (Δg_x, Δg_y and Δg_z). Here we show vertical component Δg_z maps only which reflect the strongest anomalies with respect to CAES reservoirs. We will discuss these gravity anomalies resulting from the two saturations (S_{g1} and S_{g2}) and the lower sensitivity boundary determined for the technique at all saturations, i.e., the least measurable gravity anomaly determined by the modern micro-gravimeter accuracy (3–5 μGal).

Results of Elastic Time-lapse FWI

The initial model for the time-lapse waveform inversion at each time-step is the true elastic medium model before the CAES injection. While this seems to be an overoptimistic assumption, we want to focus this part of the study on the question if the elastic FWI is capable to reconstruct structures at the resolution limit at all. Later we will also investigate the impact of different errors in the baseline model on the elastic time-lapse FWI results. To reduce the nonlinearity of the multiparameter inversion problem a sequential frequency FWI approach is applied. Therefore, Butterworth-lowpass filters are applied to the source wavelet and field data with corner frequencies of 20 and 40 Hz, respectively. To reduce the influence of multiple reflections, exponential time damping (Brossier et al.2009) of the time-lapse data after the first arrivals are applied. The first arrivals are automatically picked with a STA/LTA picker for the initial model. The detailed elastic FWI workflow is described in Table 1. All material parameters are simultaneously updated.

Table 1: FWI workflow: corner frequencies (f_c) of Butterworth-lowpass filter for the sequential frequency inversion and time-damping coefficients

f_c [Hz]	**Time-damping coefficients [1/s]**
20	100, 50, 5, 1
40	100, 50, 5, 1

al Hagrey *et al.*

al Hagrey *et al.* *SpringerPlus* 2014 **3**:267, doi:10.1186/2193-1801-3-267

The inversion results of the noise-free and noisy seismic time-lapse data (Figure 6) can be compared with the true changes of V_p, V_s and bulk density d_b in Figure 3. For the noise-free data (Figure 6, top panel) the shape of the CAES plume, the gradient in the seismic velocities and density could be reconstructed well. The magnitudes of the different material parameters also agree with the true model. The resolution limit of the elastic FWI is roughly one quarter to half of the minimum seismic wavelength. Using a maximum frequency of

60 Hz with a S-wave velocity of 2074 m/s at reservoir level leads to a minimum seismic wavelength of 34 m and therefore a resolution limit of 9–17 m. This coincides with the FD model discretization errors of 10 m at the layer interfaces visible in the FWI results. Using the noisy time-lapse data (Figure 6, bottom panel) the resolution of ΔV_p and Δd models is also quite good, beside some minor artefacts. Only the ΔV_s model is affected by the Gaussian noise. Further tests (not shown here) with a S/N = 50 (2% noise) lead to even more dominant noise artefacts, but the gas plume is still visible. At S/N = 25 (4% noise) the gas plume vanishes within the noise artefacts.

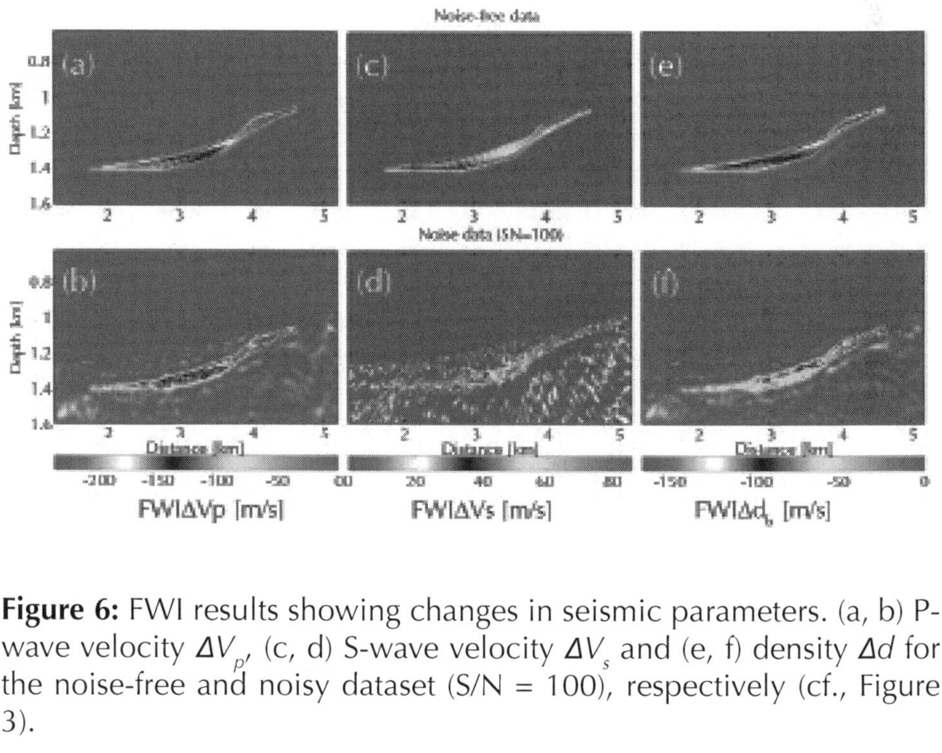

Figure 6: FWI results showing changes in seismic parameters. (a, b) P-wave velocity ΔV_p, (c, d) S-wave velocity ΔV_s and (e, f) density Δd for the noise-free and noisy dataset (S/N = 100), respectively (cf., Figure 3).

Gas Quantification by Elastic Time-lapse FWI

In most cases gas saturations are estimated from the P-wave velocity of elastic FWI results (Queisser and Singh 2012) by inverting the Gassmann equation (10). This involves the estimation of a lot more or less unknown petrophysical parameters. Because the elastic multiparameter FWI delivers a density model, the inversion of equation

(15) seems to be more appropriate. By assuming that the brine and gas densities remain constant, the following equation can be derived from equation (15) for the CAES saturation:

$$S_g^{FWI} = \frac{d_t - d_0}{\Phi(d_g - d_{br})} \tag{19}$$

depending only on the change of the bulk density between time t (d_t) and the baseline model (d_0), the porosity, as well as brine and gas density. The correct values for porosity and bulk density of the baseline model in the storage formation are used. In Figure 7, the estimated gas saturation changes from the elastic FWI density model (b, c) is compared with the true saturation (a) for the noise-free (b, d) and noisy data (c, e). Using the noise-free data, the extension of the CAES phase and the CAES saturation changes are well recovered. Due to the finite frequency content of the source wavelet, the sharp boundaries of the CAES plume cannot be resolved. Therefore, fictitious large absolute gas saturation errors of about ±10–15%, locally up to 30%, occur at the boundaries (Figure 7d, e). The average error within the reservoir approaches about 5%. With the density model from the noisy FWI result absolute saturation errors increase to 30–60% at the plume boundaries and 5–30% within the plume. Noise artefacts outside the plume can lead to errors of up to 20%. For larger S/N-ratios estimated gas saturations become unrealistically larger than 100%.

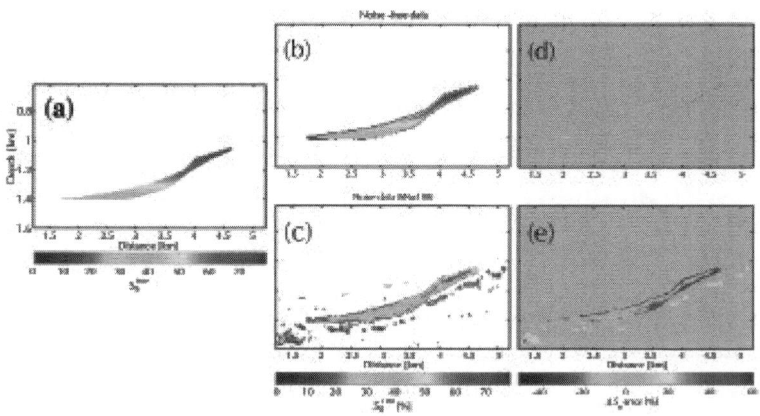

Figure 7: Gas saturation after the CAES injection. (a) True saturations S_g^{true},(b,c)S_g^{FWI} estimated from the density model of the FWI result, and (d,

e) absolute error for the noise-free data and the noisy data (S/N = 100), respectively.

Impact of Errors in the Baseline Model on the Elastic Time-lapse FWI Results

So far we only investigated the resolution of elastic time-lapse FWI when the baseline model is perfectly known. Figure 8 shows a workflow for the estimation of different baseline models with different kinds of errors. In the first step smooth versions of the true baseline seismic velocity models are generated (Figure 8a), while the density model is assumed to be constant.

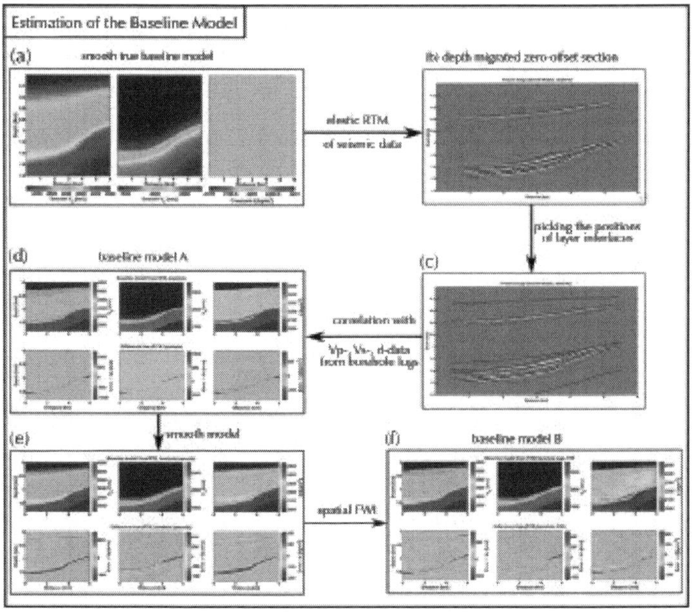

Figure 8: Workflow for the estimation of baseline models with different kinds of errors. (a) Smooth true baseline model, (b) depth migrated zero-offset section, (c) layer interfaces, (d) baseline model A, (e)smooth baseline model A, and (f) smooth baseline model B.

With the macro velocity models a pre-stack RTM can be applied to the seismic data to reconstruct the positions of the layer interfaces (Figure 8b). In the next step the interfaces are manually picked (Figure

8c) in the zero-offset section and information about the V_p, V_s and d_b model at a borehole location are used to fill the layers with material parameters (Figure 8d, baseline model A). A correct localization of the layer interfaces can be quite complicated, especially in the gas storage layer. As a result velocity and density errors of up to 1000 m/s and 300 kg/m^3, respectively, are introduced in the baseline model, even though the medium properties within the layers are lateral homogenous. Beside erroneous multiple reflections within the storage formation due to wrong interface positions, a different diffraction pattern due to changes of model discretization errors on the Cartesian FD-grid become a problem. The elastic time-lapse FWI result for baseline model A is compared in Figure 9d-f with the result for the perfect baseline model (Figure 9a-c). While the position and shape of the gas plume can be reconstructed at least to some extent, the distribution of the material properties within the gas plume are systematically overestimated. To improve the quality of baseline model A, it is smoothed (Figure 8e) with a subsequent application of spatial FWI to the baseline data (Figure 8f, baseline model B). This additional spatial FWI step reduces the errors in the velocity models at the layer interfaces up to a factor two, but the errors are still quite large with ±500 m/s for V_p, ±400 m/s for V_s, and ± 200 kg/m^3 for d_b. Additionally the smoothing introduces velocity and density errors within the layers. As a result the time-lapse FWI results of baseline model B are only slightly improved (Figure 9g-i) compared to baseline model A (Figure 9d-f).

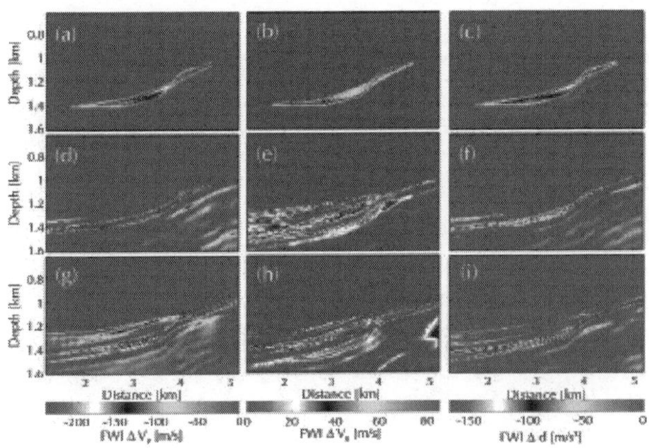

Figure 9: Influence of different baseline models on the elastic time-lapse FWI results. (a-c) Changes of V_p, V_s and density for the perfect baseline model, (d-

f) baseline model A estimated from RTM results and borehole logs, and (g-i) baseline model B estimated from RTM results, borehole logs and spatial FWI.

The evolution of the objective function *E* during the elastic time-lapse FWI is shown in Figure 10. Transitions between the different inversion stages are defined in Table 1. For the perfect baseline model without noise the objective function significantly decreases during the FWI, while none of such significant decrease is notable when data with a S/N = 100 is inverted. The number of iterations is also reduced. A comparable behaviour occurs for baseline model A and B.

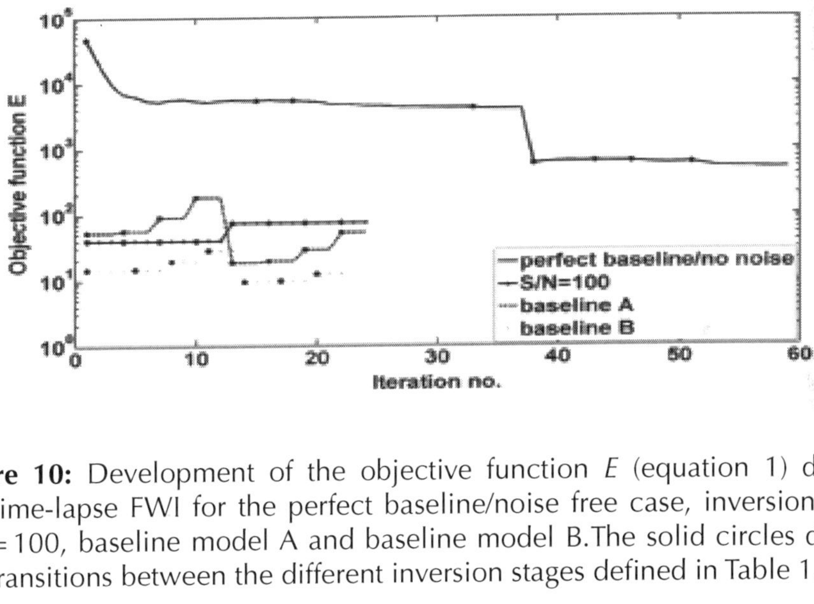

Figure 10: Development of the objective function *E* (equation 1) during the time-lapse FWI for the perfect baseline/noise free case, inversion with S/N = 100, baseline model A and baseline model B.The solid circles define the transitions between the different inversion stages defined in Table 1.

ERT Inversion Results

As mentioned before the two synthetic data sets (generated before and after adding 3% random noise) are inverted with incorporating mapping data of subsurface stratigraphy from prior surveys (e.g. seismic, borehole logs). These two apparent resistivity data sets were inverted by incorporating layer interfaces and fixing resistivity regions, respectively, both are outside the reservoir layer. ERT data inversions reconstruct directly the true subsurface resistivity tomograms including the study gas plume of downward gradual desaturation. This implies no model differencing, unlike seismic results showing the model difference before and after the gas injection. Of all differently independent inversions, every best-fitting tomogram shows least root mean square (rms)-errors of <0.5% and iteration number almost of 5, and is optimized with the L_1 norm for sharp interfaces. Also this low rms-error value is explained by the good convergence of the synthetic data sets toward the final solution. This L_1 norm yields significantly more accurate results than L_2 norm, where the actual subsurface resistivity changes abruptly at sharp target boundaries (cf. Loke et al. 2013). It is more likely to suppose that considerable subsurface information is already available during monitoring from the detailed baseline survey and any other subsequent survey. The a priori incorporation of this mapping data in ERT inversions minimizes the ambiguity of the solution and enhances the resolution of results. Here each data set inversion is constrained by incorporating resistivity regions and interfaces outside the reservoir, respectively.

Moreover, we calculated the model resistivity difference (residual) relative to the input (true) model to quantitatively evaluate the reliability of the reconstructed tomograms. This residual anomaly is a measure for the reliability of ERT inversions. It expresses quantitatively the resolution of the applied technique for both of the spatial mapping capability and the recovering resistivity amplitude. An accurate resistivity amplitude is essential for precise quantifying the injected gas phase saturation using Archie equation (17), see next section. This residual ($\Delta\rho$) between the corresponding pixels of the input or true (ρ_{true}) and output (ρ_{ERT}) 2D model is calculated by:

$$\Delta\rho_r = \frac{\rho_{true} - \rho_{ERT}}{\rho_{true}} \% \tag{20}$$

Figure 11 shows only the best-fitting tomograms reconstructed together with their corresponding residuals for the four cases of applying synthetic data before and after adding 3% random noise errors, and constrained inversions of incorporating ρ regions and boundaries outside the reservoir layer.

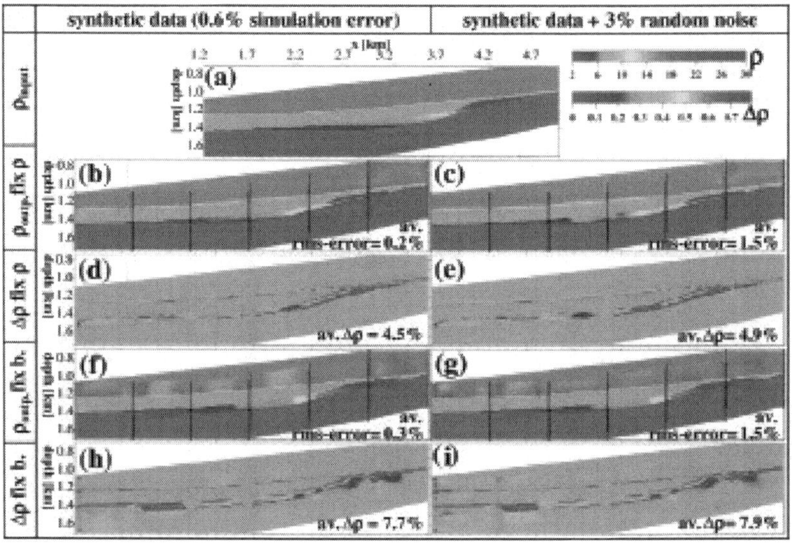

Figure 11: ERT modelling results. (a) Input or true (ρ_{true}) models, (b, c, f, g) output resistivity (ρ_{ERT}) tomograms with rms-error, and (d, e, h, i) corresponding residual $\Delta\rho$ models with their average values. Inversion is conducted using constrains of incorporating ρ region (b-e) and boundaries (f-i). These inversions are conducted for synthetic borehole ERT data (contaminated with 0.6% simulation errors) before and after adding 3% random noise error. The low residuals (average <9%) reflect well the advantage of constraining of ERT inversions by the seismic mapping results for an accurate reconstruction of the gas phase of downward gradual desaturation. The vertical black dots mark the borehole electrodes and the continuous lines mark the interfaces.

Obviously, the ERT inversion results of reconstructed tomograms visualize well the storage targets, particularly the gas plume of a downward gradual desaturation within the brine reservoir for all studied four cases (Figure 11). It is clear that constrained inversion models incorporating resistivity regions are better resolved than these incorporating boundaries. Also the addition of 3% random error to the synthetic data sets increases the misfit of rms-error values (between input and output response) by a factor of 5–8 but slightly decreases the mapping resolution as reflected by low residual $\Delta\rho$ rise from 4.5 and 7.8% to 4.9 and 7.9%, respectively. Loke et al. (2013) found that the model inverted with the L_1-norm is less sensitive to random noise compared with the L_2-norm.

Resulting residual $\Delta\rho_r$ tomograms as a measure for resolution confirm generally the good mapping capability of the applied constrained ERT technique, where all average $\Delta\rho_r$ have low values of 4.5–7.9% with minor deviations (Figure 11). This $\Delta\rho_r$ distribution shows that the resulting model reliability (inverse of $\Delta\rho_r$) is least for the noisy data set inverted by incorporating boundaries and best for the data set without adding random noise inverted with fixing ρ regions. An accurate investigation of $\Delta\rho_r$ tomograms show that the resolution suppression due to the addition of 3% random errors to the generated data (average $\Delta\rho_r$ increases from 4.5 to 4.9 for fixing ρ region, and from 7.7 to 7.9 for incorporating boundaries) is lower than that resulting from incorporating boundaries instead of fixing ρ regions (average declines from 4.5 to 7.7 for data without random noise, and from 4.9 to 7.9% for noisy data).

Constrained inversion tomograms (using any available subsurface data as an a priori information in the ERT inversion) show better resolution (inverse of $\Delta\rho_r$, Figure 11) than their corresponding unconstrained or even partly constrained inversion tomograms (not shown here, see al Hagrey et al.2013). However, these residual $\Delta\rho$ maps still reflect the common smearing effects and artifacts of varying degrees of the ERT technique. This negative effect is particularly visible within the thin storage formation ($\Delta\rho_r$ up to ±20%) with spatially varying gas saturation (i.e., resistivity amplitude). On the other hand the non-varying (homogeneous) formations above (cap rock) and below (aquitard) the storage formation show a $\Delta\rho_r$ of almost ±0. Obviously almost all inversion uncertainties are related to the monitored gas phase within the host reservoir. Thus they deliver an error estimate of the injected gas quantities monitored by this constrained inversion technique.In conclusion, the ERT technique with permanently installed borehole electrodes aims at mapping, monitoring and quantifying the gas volume injected into the saline aquifer at any time. Obviously, the resulting tomograms (Figure 11) fulfil well the spatial mapping and monitoring purposes. This permanent electrode installation helps to maximize the reliability of monitoring data. Modelled anomalies minimizes the background effect and thus maximizes the time-varying response caused here by the injected gas quantity. The residuals (Figure 11) assess and prove the high reliability of the results including the quantification capability for the resistivity amplitude. These highly reliable resistivity amplitudes motivated us to derive the gas saturations (see next section).

Gas Quantification by ERT

The gas phase saturation S_g or S_g^{ERT} in a partially saturated reservoir medium is driven indirectly by applying the resistivity amplitudes recovered from ERT models (Figure 11) in the Archie equation rearranged as:

$$S_g^{ERT} = 1 - \left(\frac{a\rho_{br}}{\rho \Phi^m}\right)^{1/n} \tag{21}$$

Here, we assume no interaction between the three reservoir phases (solid matrix, brine and gas). Using this equation, saturation models of gas phase (S_g^{ERT}) within the storage reservoir are calculated and plotted together with the corresponding input (true, S_g^{true}) models and their absolute difference ($\Delta S_g = S_g^{true} - S_g^{ERT}$) in Figure 12 for the four applied cases described before. It is clear that the estimated saturation here contains the uncertainty of the other parameters of *a,m,n,*Φ and ρ_{br} included in this equation. Unlike this, the equation of the resistivity index (the resistivity ratio after and before the injection, e.g., Nakatsuka et al. 2010) eliminates parameters *a,m* and Φ, whereas parameters ρ_{br} and *n* are assumed to stay unchanged with time. However, this index equation includes the uncertainties of both pre- and post-injection ρ models unlike our applied equation (21) which includes the uncertainty of the post injection model only. In the porous sandstone of the Rhaet reservoir, we considered the common values of 1, 2 and 2 for the constants *a,m* and *n*, respectively (Archie 1942), 0.2 for Φ, and 0.08 Ωm for ρ_{br} (as mentioned before).

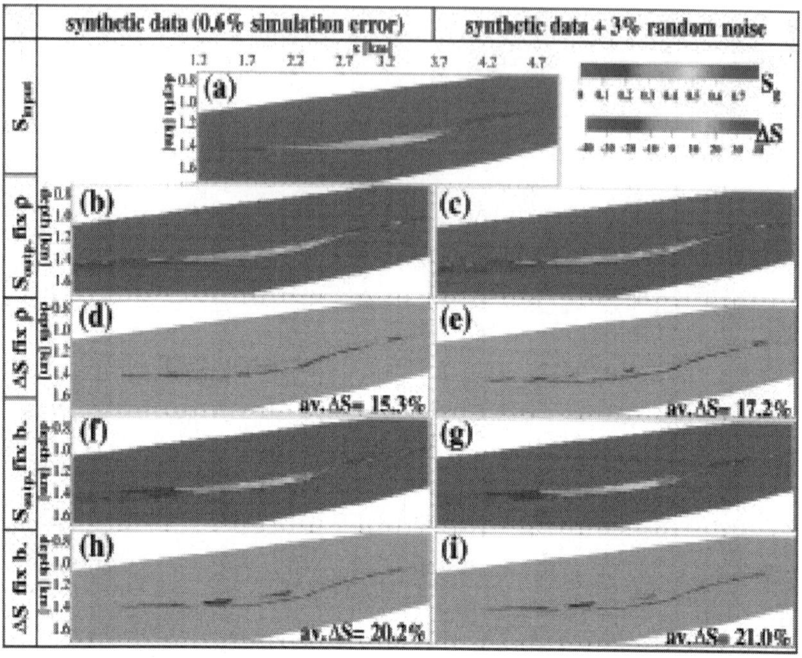

Figure 12: Results of gas phase saturations in the storage formation derived from ERT models (of Figure11). (a) True simulation (S_g^{input}), (b, c, f, g) saturations (S_g^{ERT}), and (d, e, h, i) corresponding differences (ΔS_g) with their average values. The gas phase is recovered satisfactorily everywhere as reflected by ΔS_g values (≤21%). The continuous lines mark reservoir boundaries.

An investigation of resulting models shows that the gradual gas desaturations (S_g^{ERT}) with depth are well mapped within the thin storage formation. The results in Figure 12 show a satisfactory similarity between input (true) and corresponding output (reconstructed) models. The average difference ΔS_g relative to S_g^{input} (Figure 12) approaches 15–21% for all studied four cases of Figure 11. This small difference confirms again the satisfactory reliability of the results. The gas distribution within the Rhaet reservoir formation shows a good similarity particularly in the region 1–1.4 km (corresponding to the real distribution) and is insignificant outside this region where its amplitude is below the average (ΔS_g). Most recovered S_g^{ERT} values are lower than their corresponding input ones. This may be related to the smearing effects of the technique. This smearing influences negatively the modelled resistivity amplitude and reduces the resistivity high of the gas plume sandwiched between the two resistivity lows of caprock and aquitard, respectively. High ΔS_g values are concentrated mainly at reservoir interfaces and may be related to discretization errors. At the Cranfield site, Mississipi, Carrigan et al. (2013) found that CO_2 saturations measured in monitoring wells are higher than the ERT-derived saturations although both show good spatial correlations. They added that ERT provides an integrated response from large volume, whereas gas sensors (dm penetration) provide point measurements and are sensitive to conditions near the well.

3D Gravity Modelling Results

Results of the 3D gravity modelling technique using IGMAS+ program show a high sensitivity to the applied plume scenarios of gas phase injected in the pore Rhaet reservoir of the North German Basin (Figure 13). The negative anomaly amplitude of the vertical gravity component after subtraction the background before gas injections (Δg_z) increases with increasing the gas saturation S_g causing the mass deficit. The two saturations of Sg_1 (60–27%) and S_{g2} (80–36%), gradually decreasing with depth, show negative anomalies Δg_{z1} and Δg_{z2}, respectively, of similar shape, i.e., visually both are hardly distinguishable (Figure 13a, b). The absolute Δg_{z1} amplitude for S_{g1} (134 μGal) is less than that for S_{g2} for Sg_2 (178 μGal). The difference in Δg_z amplitude (i.e., double difference) between both saturations is visualized well in figure (Figure 13c) and is far higher (>10 times) than the measurable accuracy of modern micro-gravimeters (±3–5 μGal). Fewer saturation changes are verified systematically at the whole range of saturations. We found that 1% change in saturations ΔS_g yield 2.2 μGal change in Δg_z for our study gas plume. This implies that the technique can monitor a saturation change ΔS_g down to 2.5–3% for the whole saturation range. This ΔS_g range results in a Δg_z changes above 5 μGal. Obviously, time-lapse data do not require many corrections (e.g., free-air, Bouguer, terrain) but temporal shallow changes (e.g. fluctuations of the water table) highly affect the gravity readings. Our gravity modeling of a water table at 10 m depth below the study site yields a measurable 5 μGal anomaly (micro-gravimeter accuracy) already for 0.5 m fluctuations only. Therefore, such fluctuations should be observed in wells to remove their effects from gravity readings. In conclusion, the 3D gravity modelling technique applied here is able to monitor time-lapses of gas saturations of S_{g1} and S_{g2} as well as any saturation changes (ΔS_g) down to ±3% within the whole range of saturation.

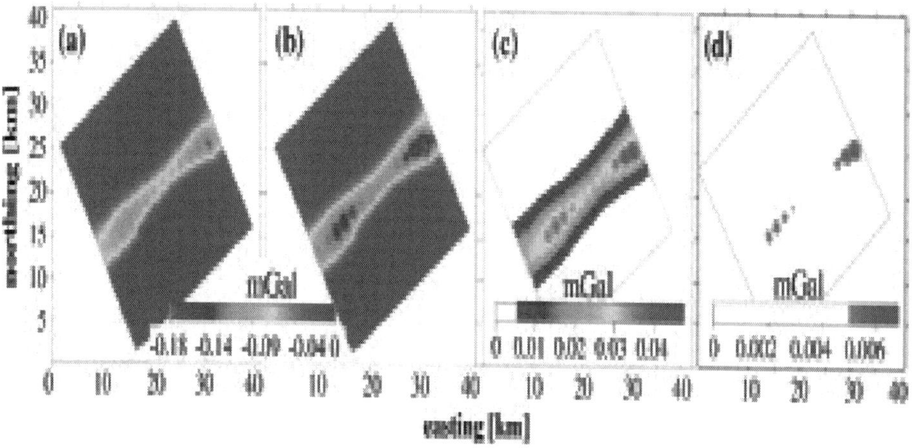

Figure 13: 3D forward gravity modelling of gas storages in the anticline limb of Rhaet sandstone formation below the study site. (a) Vertical gravity anomalies Δg_z (relative to background) for the gas reservoir with saturations S_{g1}, (b) Δg_z for S_{g2}, (c) their anomaly difference, and (d) least anomaly difference measurable by micro-gravimeters of ≥5 μGal resulting from a saturation change ΔS_g of ±3%. Only the measurable anomaly part is visualized in c-d by blanking colour bar below 5 μGal (immeasurable limit).

DISCUSSION AND CONCLUSIONS

Mitigation of anthropogenic greenhouse gases demand developments of renewable energy resources. However, most renewable energy sources are intermittent and therefore need buffer storage to match public power supply and demand. In geologic storages, replacing pore brine with compressed gas energy within the reservoir formation causes changes in elastic and electric properties and justify applications of integrative geophysical methods for monitoring this energy storage. We apply here elastic FWI, ERT and gravity techniques to map and quantify a thin gas plume of gradual downward desaturation injected in a deep brine aquifer. These aimed tasks are real challenges for any singly applied geophysical monitoring technique.

In this numerical study we simulate a nearly realistic storage scenario by considering: (1) the study site is chosen inside the North German Basin of favourable conditions for energy storages, (2) a storage model scenario is parameterized by real (published) data for this basin, (3) the gas phase plume is simulated with downward gradual desaturation similar to realistic cases, and (4) a common random noise level is added to the synthetic data to robust the technique for real field applications. The aimed resolution is enhanced by applying: (1) an integrative approach of geophysical methods, (2) an optimized approach for data acquisitions with a data coverage constraining well the inversion model and maximizing the resolution, and (3) constrained inversions to minimize interpretation ambiguities (by a priori use of available data, e.g. seismic, logs).

Unlike classical travel-time based tomographic approaches, the elastic FWI is capable to map the extension of the thin gas plume of downward gradual desaturation using only reflection seismic data, if a very accurate background model for the seismic velocities and density can be estimated before the gas injection. Additionally the elastic FWI recovers the changes of isotropic elastic material parameters and density due to the gas injection and subsequent partial drainage of the aquifer. By using an appropriate rock model, changes of the gas saturation can be deduced from the elastic FWI results with an accuracy of 5–30% within the aquifer, depending on the amount of noise (S/N-ratio > 100) present in the data. Due to the finite frequency content of the source signal larger saturation errors up to 60% can occur at the boundaries of the gas plume. Density inversion artefacts outside the aquifer due to noise can lead to fictitious estimates of saturation variations with local maxima of 20%. For a S/N-ratio of 50 the shape of the gas plume is still visible, but estimations of the gas saturation become highly erroneous and a S/N-ratio of 25 seem to be the detection limit for the gas plume. Errors in the elastic baseline model has a very substantial impact on the quality of the elastic time-lapse FWI results. Picking errors in layer interfaces and inaccurate material parameters within the layers lead to results with an approximately correct shape and position of the gas plume, but overestimated wrong elastic material parameters within the gas plume. Therefore, the estimation of accurate elastic baseline models for a successful elastic multiparameter FWI, and a subsequent calculation of the gas saturation distribution, is the greatest challenge for real field applications. Smooth macro-velocity models based on Common-Reflection Surface (CRS) stacking (Mann 2002) and Normal-Incidence Points (NIP)-wave tomography (Duveneck 2004) for P- and SH-wave data combined with the intensive use of prior information from borehole logs seem to be the most promising approach.

The applied constrained ERT inversion technique (taking use from previous seismic mapping and well logs) is also able to accurately map storage targets (caprock, reservoir with thin gas and aquitard) in all four applied cases (resulting from inverting noise-free and noisy data by incorporating layer boundaries and resistivity regions, respectively). The technique can even recover the gas plume of downward gradual desaturation with a good resolution. Also inversion models constrained by incorporating resistivity regions are better resolved than these constrained by incorporating boundaries, both are applied outside the reservoir layer. The thin resistive gas plume is sandwiched between the two conductive layers of the overlying caprock and the underlying aquitard. Based on the equivalence principle, the resistance (ρ*h, ρ = resistivity, h = thickness) of this thin resistive plume can hardly be resolved into ρ and h. This normally results in smearing with blurred boundaries and larger volume relative to the input model. These common ERT limitations are minimized here by applying the constrained inversion approach taking use of any available subsurface data. Uncertainties in mapping structures and quantifying the resistivity amplitudes are relatively low reflecting the high reliability of the reconstructed results.

Notably we could quantify reliably the gas saturations indirectly from the density and resistivity models resulting from the inversion by applying common petrophysical equations. The saturation results deduced from ERT technique fit well their corresponding values derived from elastic FWI. Both show reasonable absolute average differences (<20%) relative to the background. However, such results should be cautiously treated, where their validity and uncertainty should be studied in real field data.

The use of synthetic data contaminated with random error may reflect the real world of data. Obviously, adding random noise (typically 3%) to the synthetic ERT data increases the rms-error values by a factor of 5–9 but slightly decreases the mapping resolution. Our results here are in accordance with that obtained by al Hagrey (2012a) for ERT applications in CCS modelling. Using modelling codes (as applied here) and adding a random noise in an ascending order (1, 2, and 5% levels) to the synthetic data sets generally increases the rms-errors by a factor of 2 to 9 but slightly decreases the mapping capability of ERT technique. Ramirez et al. (2005) obtained similar results and concluded that the effect of the random error in ERT is insignificant for anomalies of a large size and magnitude.

Obviously the elastic FWI and ERT modelling using 2.5D codes has been conducted along a 2D section of the geological 3D model applied in the modelling simulation. This 2D model simplification is fully justified by the evidence that this 2D section cuts the main (storage) structure (gentle anticline within almost horizontal layering) along its main strike.

In conclusion results reveal the capability of our applied integrative geophysical approach to resolve the CAES targets and to quantify intrinsic property changes of the injected gas saturation in the reservoir. Constrained inversion models of elastic FWI and ERT are even able to recover well the gradual desaturation with depth. The accurately mapped spatial (seismic and electric) parameters are applied in their respective petrophysical equation to yield precise quantifications of gas saturations from each technique independently. Both resulting saturation models are in accordance with each other and with the input (true) saturation model. A joint elastic FWI and ERT inversion has a high potential to improve the applicability of the approach (e.g. Karaoulis et al. 2012). In a numerical study of a moving gas front within a reservoir, the joint inversion of seismic and electric time-lapse data sets reduces the presence of artifacts, and can retrieve the shape and estimate parameter better than their individual (unconstrained) inversions. For time-lapse data ERT uses permanently installed borehole electrodes, whereas seismic data needs to be repeated such that the source and receiver positions could not be the same and therefore more uncertainties can occur.

Moreover, the applied 3D gravity technique shows high sensitivity to the mass deficit resulting from the storage of the gas phase. The vertical gravity component can resolve saturations and saturation changes down to ±3% assuming that the data is corrected for temporal fluctuation effects of the groundwater table.

AUTHORS' CONTRIBUTIONS

The authors, viz SAH, DK and WR with the consultation of each other carried out this work and drafted the manuscript together. All authors read and approved the final manuscript.

ACKNOWLEDGEMENT

We thank D. De Nil for fruitful discussions, D. Wehner, M. Merz, B. Weise for computer work, the Editor and three anonymous reviewers for their constructive comments to improve the paper. Special thanks go to H.-J. Götze and S. Schmidt for supplying the license of the software IGMAS+ used for gravity modelling. This study has been carried out within the framework of research project "ANGUS+" funded by the German Federal Ministry of Education and Research (BMBF).

REFERENCES

1. al Hagrey SA (2012a) 2D optimized electrode arrays for borehole resistivity tomography and CO_2sequestration modelling. Pure and Applied 169(7):1283-1292
2. al Hagrey SA (2012b) 2D model study of CO_2 plumes in saline reservoirs by borehole resistivity tomography. International Journal of Geophysics Article ID 805059:12 doi:10.1155/2011/805059
3. al Hagrey SA, Strahser M, Rabbel W (2013) Seismic and geoelectric modelling studies of parameters controlling CO_2 geostorage in saline reservoirs. International Journal of Greenhouse Gas Control 19:796-806
4. Archie GE (1942) The electrical resistivity log as an aid in determining some reservoir characteristics. Transactions of the American Institute of Mining Engineers 146:54-62
5. Arps JJ (1953) The effect of temperature on the density and electrical resistivity of sodium chloride solutions. Transactions of the American Institute of Mining Engineers, AIME 98:327-330
6. Asnaashari A, Brossier R, Garambois S, Audebert F, Thore P, Virieux J (2013) Regularized seismic full waveform inversion with prior model information. Geophysics 78(2):R25-R36

7. Asnaashari A, Brossier R, Garambois S, Audebert F, Virieux J (2013b) Target-oriented time-lapse imaging using FWI with prior model information. EAGE Expanded Abstracts, 75th EAGE Conference & Exhibition, London, We P07 06.
8. Baldschuhn R, Frisch U, Kockel F (2001) Geotektonischer Atlas von Nordwest-Deutschland und dem deutschen Nordsee-Sektor - Strukturen, Strukturentwicklung, Paläogeographie. Geol. Jb, A153, Hannover (BGR).
9. Bauer S, Beyer C, Dethlefsen F, Dietrich P, Duttmann R, Ebert M, Feeser V, Görke U, Köber R, Kolditz O, Rabbel W, Schanz T, Schäfer D, Würdemann H, Dahmke A (2013) Impacts of the use of the geological subsurface for energy storage: an investigation concept. Environmental Earth Sciences 70:3935-3943
10. Brace WF, Orange AS, Madden TR (1965) the effect of pressure on the electrical resistivity of water-saturated crystalline rocks. J Geophys Res 70(22):5669-5678
11. Brossier R (2011) Two-dimensional frequency-domain visco-elastic full waveform inversion: parallel algorithms, optimization and performance. Comput Geosci 37:444-455
12. Brossier R, Operto S, Virieux J (2009) Seismic imaging of complex onshore structures by two-dimensional elastic frequency-domain full-waveform inversion. Geophysics 74(6):WCC63-WCC76
13. Carrigan CR, Yang X, LaBrecque DJ, Larsen D, Freeman D, Ramirez AL, Daily W, Aines R, Newmark R, Friedmann J, Hovorka S (2013) Electrical resistance tomographic monitoring of CO_2 movement in deep geologic reservoirs. International Journal of Greenhouse Gas Control 18:401-408
14. Claerbout JF, Muir F (1973) Robust modeling with erratic data. Geophysics 38:826-844
15. Cohen JK, Stockwell JJW (2008) CWP/SU: seismic unix release No. 41: an open source software package for seismic research and processing. Center for Wave Phenomena, Colorado School of Mines.
16. Crotogino F, Mohmeyer K-U, Scharf R (2001) Huntorf CAES: more than 20 years of successful operation. Solution Mining Research Institute meeting, Orlando, Florida, USA.

17. Denli H, Huang L (2009) Double-difference elastic waveform tomography in the time domain, SEG technical program expanded abstracts. 2302-2306
18. Duveneck E (2004) Tomographic determination of seismic velocity models with kinematic wavefield attributes. Ph.D. thesis, Karlsruhe University, Logos Verlag Berlin.
19. Gassmann F (1951) Über die Elastizität poröser Medien. Vierteljahrsschrift der Naturforschenden Gesellschaft in Zürich 96:1-23
20. Götze H-J, Lahmeyer B (1988) Application of three-dimensional interactive modeling in gravity and magnetics. Geophysics 53(8):1096-1108
21. Graupner BJ, Li D, Bauer S (2011) The coupled simulator ECLIPSE – OpenGeoSys for the simulation of CO2 storage in saline formations. Energy Procedia 4:3794-3800
22. Hese F (2012) 3D Modellierungen und Visualisierung von Untergrundstrukturen für die Nutzung des unterirdischen Raumes in Schleswig-Holstein. Ph.D. thesis, University of Kiel, Kiel. p 156
23. Holberg O (1987) Computational aspects of the choice of operator and sampling interval for numerical differentiation in lage-scale simulation of wave phenomena. Geophys Prospect 35:629-655
24. IEA (International Energy Agency) (2013) World energy outlook special report 2013: redrawing the energy climate map. OECD/IEA, Paris.
25. Inselmann R (1985) Dichte-Messungen in der Bundesrepublik Deutschland. Erdöl-Erdgas 101(3):76-78
26. Karaoulis M, Revil A, Zhang J, Werkema DD (2012) Time-lapse joint inversion of crosswell DC resistivity and seismic data: a numerical investigation. Geophysics 77:D141-D157
27. Köhn D, De Nil D, Kurzmann A, Przebindowska A, Bohlen T (2012) On the influence of model parametrization in elastic full waveform tomography. Geophys J Int 191:325-345
28. Komatitsch D, Martin R (2007) An unsplit convolutional perfectly matched layer improved at grazing incidence for the seismic wave equation. Geophysics 72(5):SM155-SM167
29. Landau LD, Lifschitz EM (1986) Theory of elasticity. Elsevier.

30. Levander A (1988) Fourth order finite-difference P-SV seismograms. Geophysics 53(11):1425-1436
31. Loke H, Acworth I, Dahlin T (2003) A comparison of smooth and blocky inversion methods in 2D electrical imaging surveys. Explor Geophys 34:182-187
32. Loke MH, Wilkinson PB, Chambers JE (2010) Fast computation of optimized electrode arrays for 2D resistivity surveys. Comput Geosci 36:1414-1426
33. Loke MH, Dahlin T, Rucker DF (2013) Smoothness-constrained time-lapse inversion of data from 3D resistivity surveys. Near Surface 11:20 doi:10.3997/1873-0604.2013025
34. Magri F, Bayer U, Pekdeger A, Otto R, Thomsen C, Maiwald U (2009) Salty groundwater flow in the shallow and deep aquifer systems of the Schleswig-Holstein area (North German Basin). Tectonophysics 470(1–2):183-194
35. Makhijani A, Mills C, Ramana MV (2012) Renewable Minnesota: a technical and economic analysis of a 100% renewable-energy based electricity system for Minnesota. Institute for Energy and Environmental Research (IEER), http://ieer.org/resource/reports/renewable-minnesota-technical *webcite*
36. Mann J (2002) Extensions and applications of the common-reflection-surface stack method. Ph.D. Thesis, Karlsruhe University, Logos Verlag Berlin.
37. Mora P (1987) Nonlinear two-dimensional elastic inversion of multi offset seismic data. Geophysics 52:1211-1228
38. Nakatsuka Y, Xue Z, Garcia H, Matsuoka T (2010) Experimental study on CO_2 monitoring and quantification of stored CO_2 in saline formations using resistivity measurements. Int J Greenh Gas Con 4:209-216
39. Nocedal J, Wright SJ (2006) Numerical optimization. Springer, New York.
40. Noel M, Xu B (1991) Archaeological investigation by electrical resistivity tomography: a preliminary study. Geophys J Int 107:95-102
41. Oldenburg DW, Li Y (1999) Estimating depth of investigation in dc resistivity and IP surveys. Geophysics 64:403-416

42. Queisser M, Singh SC (2012) Full waveform inversion in the time lapse mode applied to CO_2 storage at Sleipner. Geophys Prospect 61:537-555
43. Ramirez AL, Nitao JJ, Hanley WG, Aines R, Glaser RE, Sengupta SK, Dyer KM, Hickling TL, Daily WD (2005) Stochastic inversion of electrical resistivity changes using a Markov Chain Monte Carlo approach. J Geophys Res 110:18 B02101, doi:10.1029/2004JB003449
44. Schlumberger (1985) Schlumberger log interpretation charts, Schlumberger Well Services. Schlumberger Limited, New York.
45. Schmidt S, Plonka C, Götze H-J, Lahmeyer B (2011) Hybrid modelling of gravity, gravity gradients and magnetic fields. Geophys Prospect 59(6):1046-1051
46. Shipp RM, Singh SC (2002) Two-dimensional full wavefield inversion of wide-aperture marine seismic streamer data. Geophys J Int 151:325-344
47. Stummer P, Maurer H, Green AG (2004) Experimental design: electrical resistivity data sets that provide optimum subsurface information. Geophysics 69:120-139
48. Succar S, Williams RH (2008) Compressed air energy storage: theory, resources, and applications for wind power. Princeton Environmental Institute Princeton University, Princeton. p 81
49. Tarantola A (1986) A strategy for non-linear inversion of seismic reflection data. Geophysics 51(10):1893-1903
50. Virieux J (1986) P-SV wave propagation in heterogeneous media: velocity-stress finite-difference method. Geophysics 51(4):889-901
51. Virieux J, Operto S (2009) An overview of full waveform inversion in exploration geophysics. Geophysics 74(6):WCC127-WCC152
52. Wilkinson PB, Meldrum PI, Chambers JE, Kuras O, Ogilvy RD (2006) Improved strategies for the automatic selection of optimised sets of electrical resistivity tomography measurement configurations. Geophys J Int 167:1119-1126

Chapter 5

Wide-Angle Refraction Experiments in the Daito Ridges Region at the Northwestern End of the Philippine Sea Plate

Azusa Nishizawa, Kentaro Kaneda, Yasutaka Katagiri and Mitsuhiro Oikawa

Hydrographic and Oceanographic Department, Japan Coast Guard, 2-5-18 Aomi, Koto-ku, Tokyo 135-0064, Japan

ABSTRACT

Three large bathymetric highs (from north to south: the Amami Plateau, the Daito Ridge, and the Oki-Daito Ridge) originating from paleo-island arcs characterize the northwestern end of the Philippine Sea plate. We obtained 10 seismic refraction and multi-channel seismic reflection profiles across and along these bathymetric highs and obtained *P* wave velocity (Vp) models of the crust and the uppermost mantle. Although there are large variations in the crustal structure throughout this region, these bathymetric highs usually have a middle crust with Vp of 6.3 to 6.8 km/s, a lower crust with Vp of 6.8 to 7.2 km/s, a Pn velocity of 7.6 to 7.8 km/s, and a total crustal thickness of 15 to 25 km. These features are similar to those of the Izu-Ogasawara (Bonin)-Mariana island arc and the Kyushu-Palau Ridge, which are immature paleo-island arcs. However, the crust at the southwestern part of the Oki-Daito Ridge contains a relatively thin middle crust and a smaller total crustal thickness compared with other ridges in this region. In addition, we identified a deep reflector beneath the ridge, with these properties indicating a different origin, such as intraplate volcanism.

BACKGROUND

The Philippine Sea plate is composed of three inactive marginal basins, the Shikoku Basin, the Parece Vela Basin, and the West Philippine Basin, as well as many bathymetric highs (Figure 1). The evolutionary histories of these basins have been mostly elucidated based on geophysical and geological data (e.g., Kobayashi et al. 1995;Okino et al. 1999;Deschamps and Lallemand 2002). The northwestern end of the Philippine Sea plate is characterized by several old bathymetric highs, known as the Daito Ridges. From north to south, these comprise the Amami Plateau, the Daito Ridge, and the Oki-Daito Ridge, with models for their evolution having been presented in various studies (e.g., Tokuyama et al. 1986;Tokuyama 2007;Ishizuka et al. 2013). The latitude at which these bathymetric highs formed was investigated by Ueda (2004) using topographic and magnetic anomaly data. Paleolatitudes of 15.1° ± 4.6°N for the Amami Plateau, 20.1° ± 8.2°N for the Daito Ridge, and 4.4° ± 11°S for the Oki-Daito Ridge were obtained, showing that these bathymetric highs initially formed south of their present positions and have since migrated northwards.

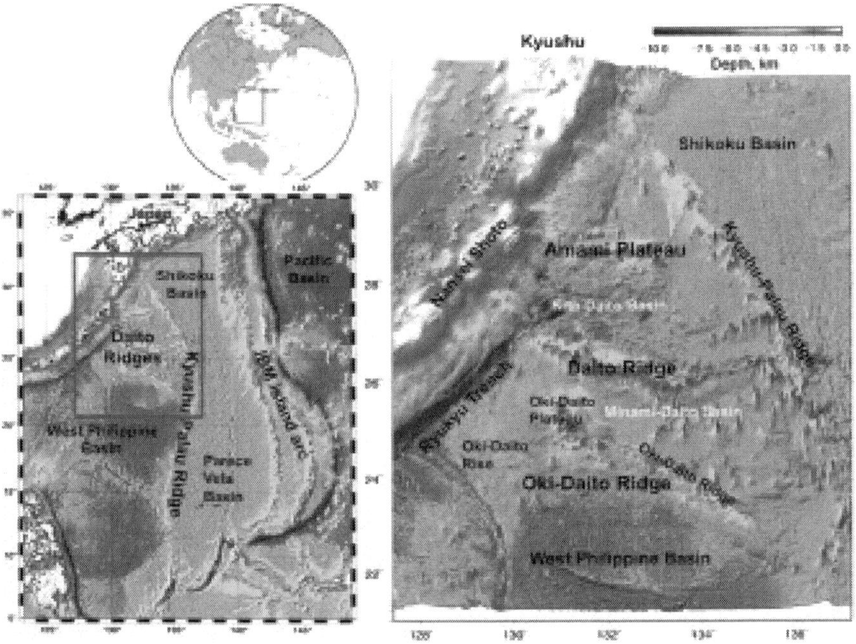

Figure 1: Tectonic map. The characteristic seafloor topography of the Philippine Sea plate on the left and detail of the Daito Ridges region (red box) magnified on the right.

The topographic highs of the Amami Plateau consist of three east-west-oriented ridges that extend for approximately 300 km, with the middle ridge being both the largest and the shallowest in depth (less than 2,000 m). The northwestern end of the plateau is currently subducting beneath the Ryukyu (Nansei-Shoto) Trench (Nishizawa et al. 2009). Igneous rocks obtained from the plateau suggest that its pedestal is Cretaceous in age, and the ridge portion situated on the pedestal formed in the Middle Eocene (Tokuyama et al. 1986;Ishizuka and Yuasa 2007). Hickey-Vargas (2005) described basalt and tonalite from the Amami Plateau that have geochemical characteristics of intra-oceanic island arc rocks and which formed at 70 to 118 Ma during the Cretaceous. Based on these data, Tokuyama (2007) concluded that the Amami Plateau was an active island arc before the Eocene.

The Daito Ridge is situated south of the Amami Plateau; its western end approaches the Ryukyu Trench, and its eastern end connects to the Kyushu-Palau Ridge. The Daito Ridge is 500-km long, east-west trending and shows an arched topography facing south. The ridge is divided into a shallow, wide, and flat bottom in the west and a narrow ridge in the east. Samples of basalt obtained from DSDP Site 445 at the center of the ridge (Figure 2a) were determined by Tokuyama et al. (1986) to be middle Eocene in age and to have originated in an island arc setting. Furthermore, Ishizuka and Yuasa (2007) estimated the age of the aphyric basalt recovered from the western part of the Daito Ridge to be 116.9 Ma, which mostly agrees with the reported ages of igneous rocks elsewhere in the Amami Plateau.

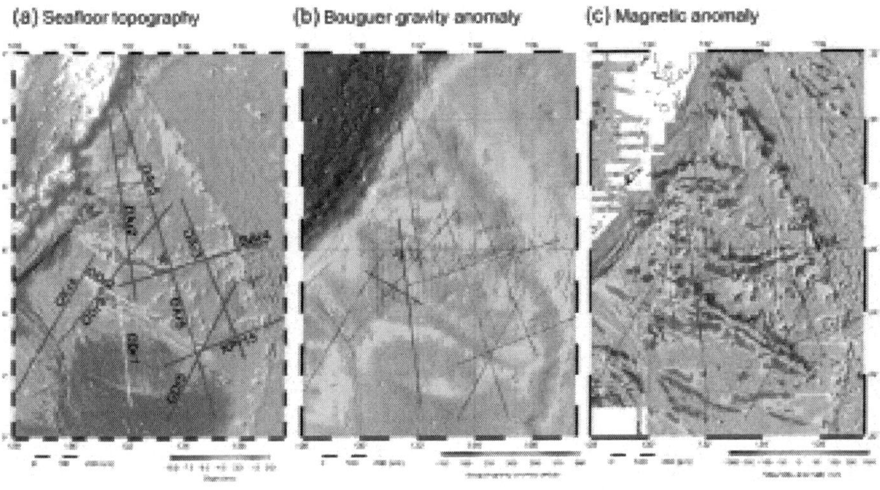

Figure 2: Positions of the seismic profiles. (a) A bathymetric map, (b) a Bouguer gravity anomaly map, and (c) a magnetic anomaly map superimposed on the shaded topographic map. The position of DSDP 445 in (a) is indicated with a star.

Further south, the Oki-Daito Ridge, in the broad sense, has a total length of approximately 700 km and is topographically divided into three parts: the Oki-Daito Rise in the southwest, the Oki-Daito Plateau in the northwest, and the Oki-Daito Ridge in the strict sense hereafter referred to simply as the Oki-Daito Ridge in the east (Figure 1). The Oki-Daito Plateau connects with the Daito Ridge at its northwestern end. The Oki-Daito Ridge is characterized by two rows of linear bathymetric highs that have a V-shaped topographic depression between them. Volcanic edifices on the Oki-Daito Ridge and the Minami-Daito Basin to the north have geochemical characteristics similar to ocean island basalt (OIB) and have an estimated age of 44 to 48 Ma (Ishizuka et al.

2013). By contrast, alkali basalt dredged at the southeastern bathymetric high on the Oki-Daito Ridge has an age of 65 Ma and formed through undetermined volcanic activity (Okino and Kato 1992). As a result, these data indicate that volcanic activity has occurred on at least two separate occasions in this region.

Kasuga et al. (1986) analyzed gravitational and geomagnetic data from the region around the Daito Ridge and the northern Oki-Daito Ridge, and reported that free air and Bouguer gravity anomalies indicate that the crust below the ridges is thick. Furthermore, magnetic anomalies suggest that these ridges have a magnetic basement with properties characteristic of intermediate and acidic rock types, combined with several intrusive rocks that have intermediate and basic rock properties (Kasuga et al.1986). They concluded that the two ridges are remnants of island arcs, on the basis of previous results from rock samples and other geological features.

Ishihara and Koda (2007) gravimetrically estimated the crustal thickness of the Philippine Sea plate using a simple four-layer model, which comprised seawater, sediments, crust, and lithospheric mantle, with respective densities of 1,030, 2,300, 2,800, and 3,300 kg/m^3. Their results show that the Amami Plateau, the Daito Ridge, and the Oki-Daito Ridge (in the broad sense) have crusts of 15- to 20-km thickness. However, they could not constrain its overall constitution (i.e., the relative proportions of the upper, middle, and lower crusts). Furthermore, there is insufficient information about densities and sediment distribution to obtain a more thorough image of the crustal structure in this region.

Various seismic refraction experiments have been conducted in this region in an attempt to reveal more of its crustal structure. Murauchi et al. (1968) revealed that the Oki-Daito Ridge has a continental-type upper crust that is up to 5 km thick and has a *P* wave velocity (Vp) of 6 km/s. Exploration conducted by Nishizawa et al. (1983) using a small number of ocean bottom seismographs (OBSs) produced similar crustal models for beneath the Amami Plateau, which indicate an upper crust up to 4 km thick with Vp of 5.4 to 5.6 km/s. However, they did not obtain the precise depth of the Moho due to a lack of seismic energy and insufficient travel time data. There was also little available information to image the entire paleo-island arc crust region of the Daito Ridges.

Recently, Calvert (2011) reviewed island arc seismic velocity models and discussed two examples as evolutionary end members: the Aleutian arc, initiated in the Eocene and not significantly affected by extension and rifting, and the Izu-Ogasawara (Bonin)-Mariana (IBM) arc-backarc system, also initiated in the Eocene but variably affected by localized extension and two episodes of arc rifting and backarc spreading. Calvert (2011) divided both arc crusts into three parts based on their Vp and crustal thickness proportions and showed that their middle and lower crusts have higher seismic velocities than those of typical continental crust. Moreover, the Aleutian middle crust has Vp of 6.5 to 7.3 km/s, which is significantly higher than that of the IBM middle crust (6.0 to 6.8 km/s). Thus, comparing the arc crustal structure of the Daito Ridges with these arc models will provide important insight into their evolution.

From 2004 to 2006, we carried out extensive wide-angle seismic and multi-channel seismic (MCS) explorations in the Daito Ridges region as part of the Japanese Continental Shelf Survey Project. This paper presents the first seismic images of the deeper structures beneath the bathymetric highs in this region and aims to contribute to future research into the early history of the Philippine Sea plate.

METHODS

The locations of our seismic profiles in relation to known bathymetric, Bouguer gravity anomaly, and magnetic anomaly data are shown in Figure 2. In this seismic experiment, we shot a tuned array of air guns with a volume of 8,040 $in.^3$ (132 L) at intervals of 200 m (90 s) for the wide-angle seismic profiles and at intervals of 50 m for the MCS profiles (480 channels, 60-folds). OBSs were deployed at an average interval of 5 km, except for ODr1, where the interval was 3 km.

We used pop-up type OBSs equipped with a three-component 4.5-Hz geophone and a hydrophone. Sensor outputs were continuously recorded on a hard disc at a sampling rate of 200 Hz with a resolution of 24 bits. Each OBS has an acoustic release unit for retrieval. The OBS locations were determined using acoustic waves from a ship with GPS navigation and direct water waves from an air gun array to the OBS.

OBS record sections were produced by 4- to 16-Hz bandpass filtering, predictive deconvolution (first zero crossing, operator length of 660 ms), and a local slant stack to enhance the signal-to-noise ratio. We read out the arrival times of the refraction and reflection signals from these record sections. Using these data, we produced Vp models by a combination of tomographic inversion and two-dimensional forward modeling. An initial velocity model was produced with the upper sedimentary layer constrained by the MCS data. First arrival data were inverted for a two-dimensional velocity model using the tomo2d tomographic inversion code of Korenaga et al. (2000). We constructed the model from shallower to deeper crust by increasing the offset limits of the input travel time data. In the velocity models derived for all profiles, the horizontal grid spacing is 0.5 km, and the vertical grid spacing gradually increases with depth according to the relation $0.05 + (0.01 \times \text{depth (km)})^{1/2}$ km.

Tomographic inversion misfits were less than 50 ms for almost all velocity models, and the resolution and reliability of our models were examined using checkerboard tests. We built a reference model by adding sinusoidal anomalies with a horizontal dimension of 10 km, a vertical dimension of 2.5 km, and a velocity perturbation of ±5% to the upper crust of our preferred final model. The vertical cells in the middle and lower crusts were 5 and 10 km, respectively, in the vertical direction and 5 km in the horizontal direction. We used ray coverage in addition to checkerboard test results to represent model reliability. Better recovery of the checkerboard pattern and higher ray density indicate higher resolution.

To utilize reflection arrivals and low-amplitude refraction signals at distant offsets, we carried out forward modeling using a two-dimensional ray-tracing algorithm (Fujie et al. 2000;Kubota et al.2009). We used the tomographic model as an initial ray-tracing model and then improved it through trial and error. Where the final result needed to be confirmed, additional tomographic inversion was conducted using the ray-tracing model as an initial model of inversion. The final velocity models in this study are the results of forward modeling, except for DAr5. The total length of DAr5 is 1,185 km, the longest line in our exploration, and thus, we did not examine the model by forward modeling due to time limitations. In the last step, two-dimensional synthetic seismograms were calculated by the finite difference method, using the E3D code of Larsen and Schultz (1995), and compared with the observed amplitude variation in the record sections.

RESULTS

The calculated Vp models discussed below are presented from north (the Amami Plateau) to south (the Oki-Daito Ridge). The Amami Plateau was transected by two north-south-oriented seismic lines (DAr2 and DAr5), whereas six lines were run across and along the Daito Ridge (ODr1, DAr2, ODr9, DAr5, DAr3, and DAr4). We also report Vp models for these north-south profiles, except for line DAr4, which does not transect the main body of the Daito Ridge. Seven seismic lines were shot across and along the Oki-Daito Ridge. However, since line ODr2 was too short to resolve the deeper part of the Oki-Daito Ridge, we only show results for OKr4, ODr1, and ODr9 for the northwestern part of the ridge and ODr5, DAr5, and KPr15 for the southwestern part.

Following Calvert (2011), we loosely divide our crustal models into three parts (upper, middle, and lower crusts) based on their velocities, velocity gradients, and thickness proportions, and subsequently describe their characteristics.

Amami Plateau

Figure 3 shows the Vp models along the north-south lines DAr5 and DAr2. DAr2 crosses the shallowest part of the region, where water depth only reaches approximately 1,550 m. The record section, ray diagram, and synthetic seismogram for OBS 41 on DAr2 are shown in Figure 4. The fitting error between the observed and calculated first arrivals is less than about 50 ms. The crustal models in Figure 3 are strongly heterogeneous in horizontal direction, especially in the upper crust, with Vp of about 3.0 to 6.0 km/s. A middle crust characterized by Vp of 6.0 to 6.8 km/s and a thickness of approximately 2 to 4 km is present beneath the plateau, except below the southern row of the bathymetric high along DAr2, where little material with a similar Vp exists. The middle crust in the western region of the plateau that contains greater water depths along DAr2 is relatively thin when compared with that in the eastern region along DAr5. Vp of the lower crust is 6.8 to 7.2 km/s.

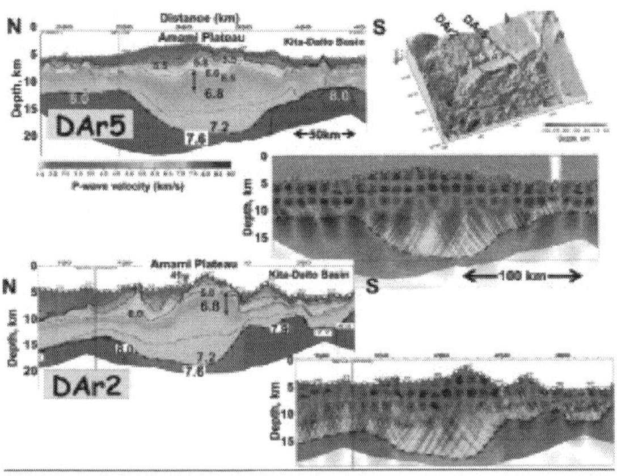

Figure 3: Vp models of the Amami Plateau for lines DAr5 and DAr2.The double-headed arrow indicates a region of middle crust with Vp of 6.0 to 6.8 km/s. The record section obtained at OBS 41 on DAr2 (blue inverted triangle) is shown in Figure 4. Checkerboard test results and ray coverage for the models are also shown. The inset in the upper right corner shows the position of the profiles on a three-dimensional bathymetric map.

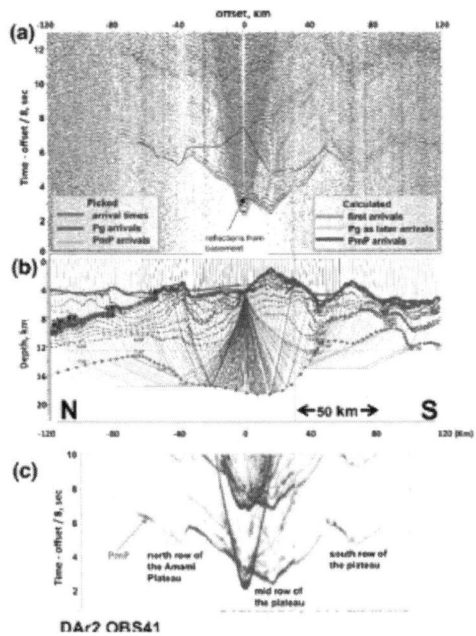

Figure 4: Observed record section, ray diagram, and synthetic seismogram. (a) Observed record section for OBS 41 of line DAr2 on the Amami Plateau.

The reduction velocity is 8 km/s. The arrival time readings and calculated travel times are shown on the record section. The calculated Pg arrivals between −30 and 30 km are artificial. (b) Ray diagram. (c)Synthetic seismogram.

PmP travel times infer a maximum depth of 19 km for the Moho discontinuity beneath the middle-plateau bathymetric high (Figure 4), whereas the crust below the southern high is the thinnest of the three rows. The Pn velocity immediately beneath the plateau is approximately 7.6 km/s (significantly lower than 8 km/s), which is verified by both inversion and forward modeling. The synthetic seismograms in Figure 4c also explain both observed travel times and amplitude variation with an offset.

Daito Ridge

Figures 5 (top left) and 6a show the crust and uppermost mantle models beneath the north-south lines across the Daito Ridge, respectively. In addition to the Vp models, the checkerboard test results of tomographic inversion and the ray coverage are also shown in Figures 5 (bottom left) and 6b, respectively. The checkerboard pattern can be recovered down to a depth of 15 km under the ridge, with the deeper part determined mainly by forward modeling. We do not present a Vp model for DAr4 along the eastern thin ridge in this paper, but we have examined its consistency at the intersections with DAr5 and DAr3. Although the upper crust Vp, the depth of the Moho, and the Pn velocities are mostly consistent at these seismic line intersections, the velocity profiles of the middle and lower crust are different. This may be due to large horizontal heterogeneities in each crust; the structural model beneath ODr1 is shown in the next section for the Oki-Daito Ridge.

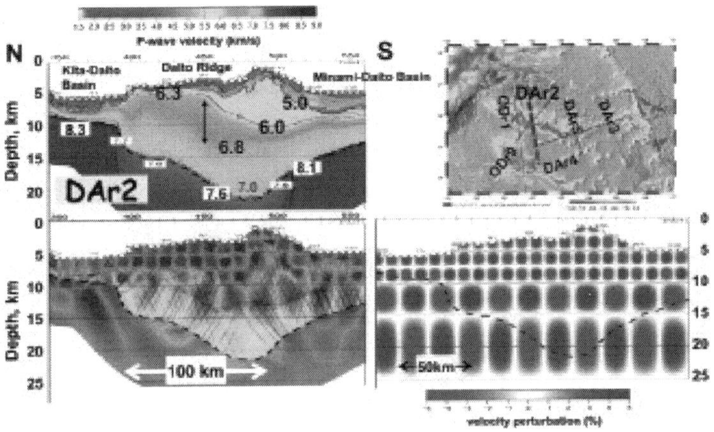

Figure 5: Vp model of the western Daito Ridge. Line DAr2 (top left), checkerboard test results and ray coverage (bottom left), and checkerboard test pattern (bottom right). The double-headed arrow indicates a region of middle crust with Vp of 6.3 to 6.8 km/s. The checkerboard pattern can be recovered to a depth of 15 km under the ridge, with the deeper part mainly determined by forward modeling. The position of the profile on the bathymetric map is shown in the top right panel.

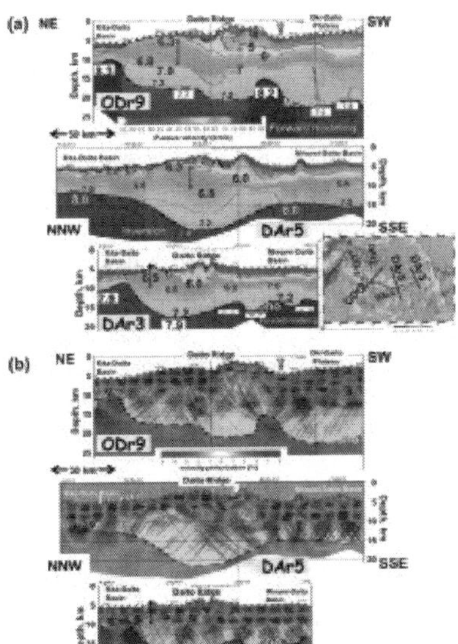

Figure 6: Vp models of the Daito Ridge and checkerboard test results and ray coverage. (a) Vp models of the Daito Ridge for lines ODr9 (top), DAr5

(middle), and DAr3 (bottom). The double-headed arrow indicates a region of middle crust with Vp of approximately 6.3 to 6.8 km/s. The inset in the lower right corner shows the locations of the profiles on the bathymetric map. (b) Checkerboard test results and ray coverage for the Vp models of the Daito Ridge for lines ODr9 (top), DAr5 (middle), and DAr3 (bottom) that are shown in (a).

Despite the large heterogeneity among the Vp models for the DAr2, ODr9, DAr5, and DAr3 lines, a common characteristic of these velocity structures is another notable difference between the northern and southern parts of the ridge. A material with $Vp \leq 6.3$ km/s and a thickness of 6 to 11 km is present in the southern region, while a material with $Vp \geq 6.3$ km/s ascends to approximately 2 km from the seafloor in the northern region.

Determining the velocity at the base of the crust can be difficult, but it is sometimes possible to estimate the velocity distribution in the entire crust effectively by using as many later phase arrivals as possible. The observed first and later arrivals are explained by forward modeling with a lower crust Vp of 7.0 to 7.2 km/s (Figure 7a, left). Vp of the uppermost mantle is 7.6 to 7.9 km/s, significantly lower than 8.0 km/s beneath the bathymetric highs, which is consistent in the northern and southern regions of the Daito Ridge. Tomographic inversion with a broad Moho transition as an initial model resulted in an uppermost mantle $Vp \leq 8$ km/s. The crustal thickness varies between 13 and 20 km along the seismic lines and is generally thinner in the northern compared to the south for DAr2 and ODr9 beneath the western ridge, which corresponds to the difference in water depth. Regarding DAr5 and DAr3 in the eastern ridge with a narrow bathymetric high, the crustal thickness is the same for the north and south regions, and the crusts are thickest around the middle point of the ridge.

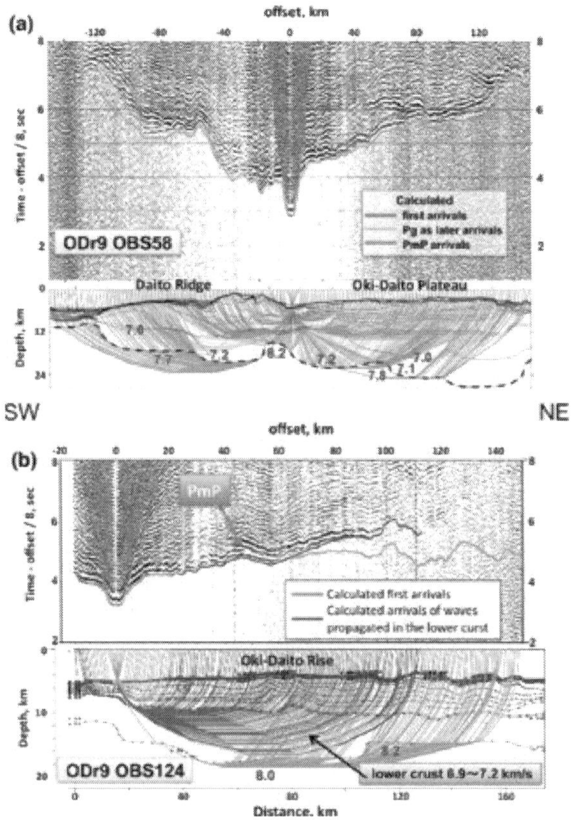

Figure 7: Examples of forward modeling using observed later phases to constrain the velocity distribution of the entire crust. (a)Observed record section (top) and ray diagram (bottom) for OBS58 of line ODr9. The position of the OBS is shown in Figure 6a, b. (b) Observed record section (top) and ray diagram (bottom) for OBS 124 of line ODr9. The position of the OBS is shown in Figure 9a, b. The reduction velocity is 8 km/s and calculated travel times are shown on the record section.

Oki-Daito Ridge

Six velocity profiles across the Oki-Daito Ridge in the broad sense are presented in this work. Three of them (DAr5, ODr5, and KPr15) intersect the bathymetric high at the southeastern Oki-Daito Ridge (Figure 8a), and the other three (OKr4, ODr9, and ODr1) transect the Oki-Daito Rise and Oki-Daito Plateau (Figure 9a). The checkerboard test results and ray diagram for each model are shown in Figures 8b and 9b.

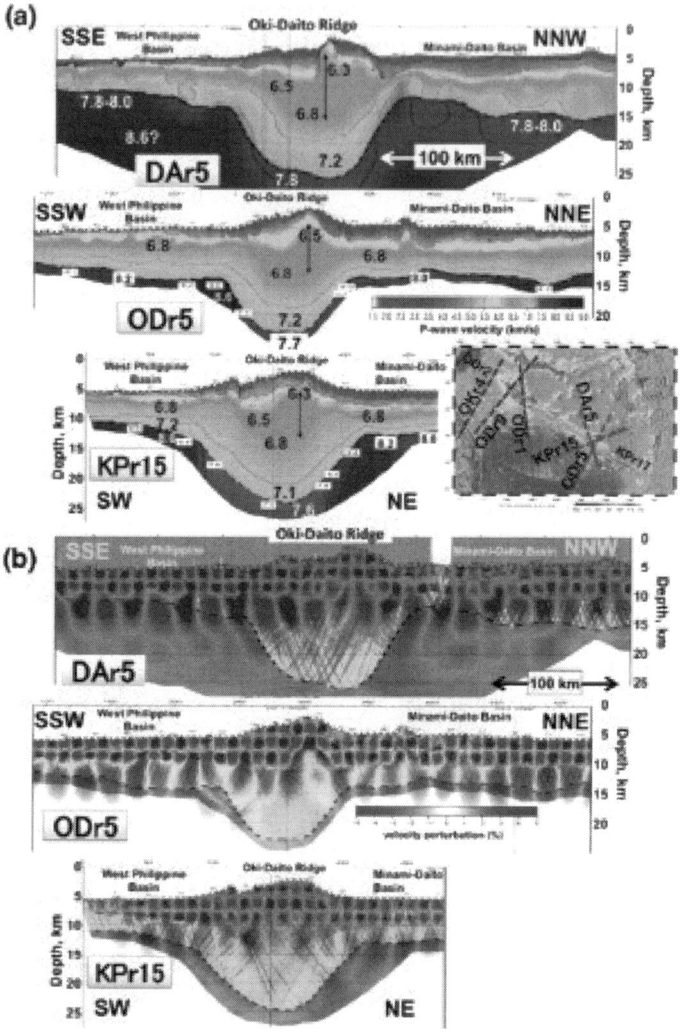

Figure 8: Vp models and checkerboard test results. (a) Vp models of the southeastern Oki-Daito Ridge for lines DAr5 (top), ODr5 (middle), and KPr15 (bottom). The double-headed arrow indicates the thickest region of the middle crust with Vp of approximately 6.3 to 6.8 km/s for each profile. The inset in the lower right corner shows the locations of the profiles (red dashed lines) on the bathymetric map. (b) Checkerboard test results and ray coverage for Vp models of the southeastern Oki-Daito Ridge for lines DAr5 (top), ODr5 (middle), and KPr15 (bottom) that are shown in (a).

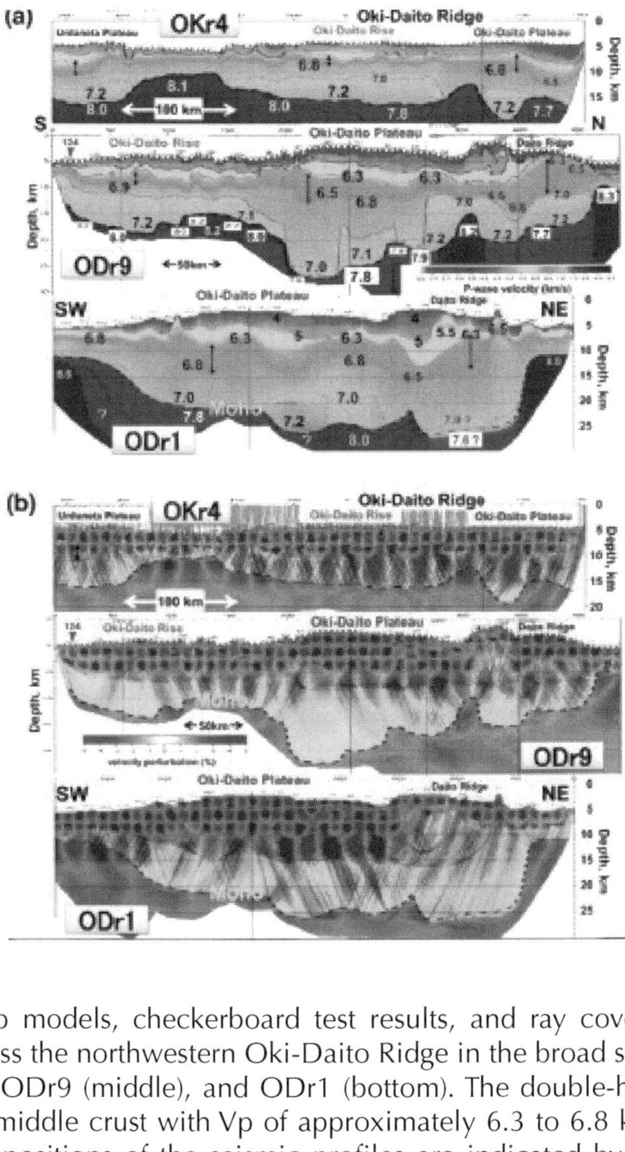

Figure 9: Vp models, checkerboard test results, and ray coverage. (a) Vp models across the northwestern Oki-Daito Ridge in the broad sense for lines OKr4 (top), ODr9 (middle), and ODr1 (bottom). The double-headed arrow indicates a middle crust with Vp of approximately 6.3 to 6.8 km/s for each profile. The positions of the seismic profiles are indicated by blue dashed lines in the lower right corner in Figure 8a. **(b)** Checkerboard test results and ray coverage for Vp models across the northwestern Oki-Daito Ridge in the broad sense for lines OKr4 (top), ODr9 (middle), and ODr1 (bottom) that are shown in **(a)**.

Figure 8a illustrates three Vp models across the two rows of linear bathymetric highs and a valley between them in the southeastern Oki-Daito Ridge. The three profiles transect a bathymetric high on the northern row, the shallowest portion in the southeastern Oki-Daito

Ridge. A common feature is the presence of a thick middle crust with Vp of 6.3 to 6.8 km/s beneath the bathymetric high. The upper crust (with Vp of approximately 3.0 to 6.3 km/s) is slightly thicker beneath the valley. The thickness of the lower crust (with Vp of 6.8 to 7.2 km/s) is 7 to 10 km, and the total crustal thickness is 20 to 30 km. Although the Pn velocity is about 8.6 km/s at the transition from the Oki-Daito Ridge and West Philippine Basin, the velocity below the ridge crest is 7.6 to 7.8 km/s. The high uppermost mantle velocity of about 8.6 km/s seems to belong to the West Philippine Basin, which is horizontally and vertically heterogeneous and has a Pn velocity ranging from approximately 8.0 to 8.6 km/s (Figure 8a, b).

Vp and the thickness of the uppermost sediment layer in the Oki-Daito Plateau are 1.8 to 2.2 km/s and about 1 km, respectively (Figure 9a). The upper crust has Vp of about 3.0 to 6.3 km/s and a thickness of 3 to 7 km. Underlying the upper crust, there is a middle crust with Vp of 6.3 to 6.8 km/s and a thickness of 3 to 5 km, and a lower crust with Vp of 6.8 to 7.2 km/s and a thickness of 10 to 15 km. The Pn velocity here is 7.7 to 7.9 km/s, and the total crustal thickness is 20 to 25 km. These characteristics are similar to those of the Daito Ridge.

Clear large-amplitude PmP signals were observed in the record sections of many OBSs installed on the plateau. Later phases with comparable large amplitude were frequently recorded at larger offsets (e.g., Figure 10b), which makes the identification of PmP difficult. We interpreted the shallower reflector as the Moho discontinuity and the deeper one as a reflector in the uppermost mantle, 5 to 10 km below the Moho discontinuity.

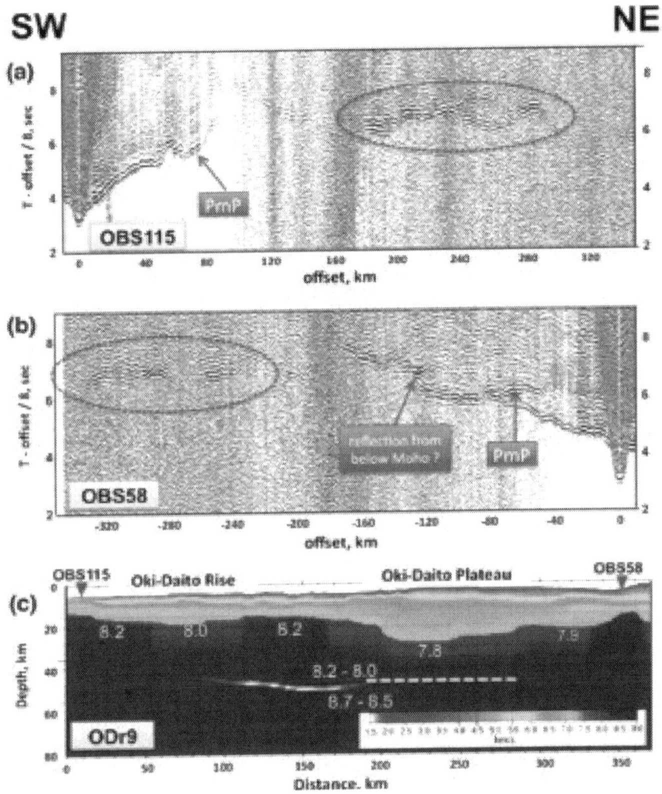

Figure 10: Vertical record sections and estimated position of a velocity discontinuity. (a, b) Vertical record sections obtained at OBS115 and OBS58 on line ODr9. Large-amplitude signals (denoted with red ellipses) are used for reflector mapping. (c) Estimated position of a velocity discontinuity in the mantle beneath the Oki-Daito Plateau and Oki-Daito Rise. The white and yellow lines were derived from mapping and forward modeling, respectively. A reflector at a depth of around 45 km could roughly explain the travel times and amplitudes of signals at offsets of over 180 km.

The middle crust of the Oki-Daito Rise is thin compared to that at other bathymetric highs in this region (Figure 9a). The Vp distribution in the crust is well-constrained, as shown in Figure 7b, and Vp at the crust bottom is estimated to be 7.2 km/s. The total maximum crustal thickness of 14 km is also thinner compared with other ridges in this region. The Pn velocity is 7.8 km/s in the northern part of the rise along OKr4 (Figure 9a, top), but a higher Pn velocity (≥8.0 km/s) was measured along ODr9 (Figure 9a, center) and the southern rise along OKr4.

OBSs deployed on the Oki-Daito Plateau and Oki-Daito Rise frequently recorded large-amplitude signals over distances of about 180 km. Figure 10 shows examples of the record sections obtained at OBS 58 and 115 on ODr9. When considering these signals as first arrivals, we were unable to construct a Vp model to account for these travel times; thus, we interpreted the travel times as reflection signals and estimated the position of the reflector using the travel time mapping method of Fujie et al. (2006). The reflector was denoted with a white line in Figure 10c. In addition to the mapping, we examined the reflector position and Vp distribution by forward modeling. Assuming a rough estimate of the mantle velocity and a reflector extending for 110 to 290 km along ODr9 (Figure 10c), we were able to explain the observed travel times with an error of less than 0.2 s (except at offset 210 to 240 km for OBS 115, where the error was 0.3 s). Finally, we compared the observed record sections with synthetic seismograms and confirmed large amplitude signals at offsets of over 180 km. That is, a velocity discontinuity at a depth of about 45 km under the plateau is proposed as one of possible interpretations for these signals. Furthermore, similar reflection signals were recorded by several OBSs at OKr4. In the same way, we estimated the position of the interface at a depth of about 50 km with a horizontal width of about 60 km beneath the Oki-Daito Rise.

DISCUSSION AND CONCLUSION

We have compiled Vp structural models from ten wide-angle seismic lines in the Daito Ridges region at the northwestern end of the Philippine Sea plate. The crustal models obtained for the Amami Plateau (Figure 3) and the Daito Ridge (Figures 5 and 6a) are similar. In both cases, there is an upper crust with Vp of 3.0 to 6.3 km/s and a middle crust with Vp of 6.3 to 6.8 km/s below the ridges, although their relative proportions vary along the seismic lines. The lower crust and upper mantle seismic velocities are 6.8 to 7.2 and 7.6 to 7.9 km/s, respectively. The maximum crustal thickness beneath the bathymetric highs reaches approximately 20 km.

The Kita-Daito Basin is situated between the Amami Plateau and Daito Ridge (Figure 1) and has a thin crust of 4 to 6 km, as deduced from its seismic structure (Nishizawa et al. 2013). The top of the middle crust of the Amami Plateau (Figure 3) tends to shallow southward of the Kita-Daito Basin, particularly in the mid-row bathymetric high along DAr2. A notable feature of the Daito Ridge crust is that a material with Vp $\leq$ 6.3 km/s and a thickness of 6 to 11 km is concentrated in the southern part, while a material with Vp $\geq$ 6.3 km/s ascends closer to the seafloor toward the Kita-Daito Basin in the north. The middle crust beneath both the Amami Plateau and the Daito Ridge shallows towards the Kita-Daito Basin, which may be the result of extension between the two bathymetric highs. However, instead of an ascending middle crust, Canales et al. (2008) report that a deeper crust consisting predominantly of gabbro was exhumed at several oceanic core complexes, which are large-offset normal fault systems commonly referred to as oceanic detachment faults on the Mid-Atlantic Ridge.

A much thinner crust, with a thickness of approximately 3 km and a high Pn velocity of 8.3 km/s, occurs at the transition from the Daito Ridge to the Kita-Daito Basin. This crust shows the same characteristics as that found at the transition between the Kyushu-Palau Ridge and the Shikoku and Parece Vela Basins, where backarc spreading had occurred (Nishizawa et al. 2012). The Vp model of the Kita-Daito Basin suggests the presence of a thinner crust, similar to the backarc basin oceanic crust in the Shikoku and Parece Vela Basins, as revealed by Nishizawa et al. (2011 2013). These features lend credence to the interpretation of rifting and seafloor spreading between the Amami Plateau and Daito Ridge, which is supported further by roughly east-west aligned magnetic anomalies in the Kita-Daito Basin (Figure 2c) that imply seafloor spreading in the north-south direction. By contrast, Kasuga et al. (1986) interpreted these lineations to be related to topographic relief. Moreover, although Tokuyama et al. (1986) proposed that the backarc spreading of the Kita-Daito Basin was caused by subduction at the southern edge of the Daito Ridge, our seismic profiles do not contain any distinctive features indicative of subduction.

Individually, Vp models for the three sub-regions of the Oki-Daito Ridge in the broad sense reveal different features (Figures 8a and 9a). The Oki-Daito Plateau has the thickest crust (20 to 25 km) of these sub-regions (Figure 9a). This comprises an upper crust with Vp of approximately 3.0 to 6.3 km/s, a middle crust with Vp of 6.3 to 6.8 km/s, and a lower crust with Vp of 6.8 to 7.2 km/s, which is a similar structure to that of the Daito Ridge crust. In contrast, the Oki-Daito Rise in the southwestern part of the ridge has a relatively thin middle crust, reaching a maximum total crustal thickness of 14 km, which is different from other bathymetric highs in the Daito Ridges as well as from typical oceanic crust.

A deep reflector was identified at approximately 45 km below the sea floor between the Oki-Daito Rise and Plateau (Figure 10). The detection of such deep reflectors requires a sufficiently long seismic line as well as closely placed shots and receivers, although we do not presently have sufficient data to discern the spatial distribution of these reflectors in the Daito Ridges region. In addition, most seismic refraction lines performed in previous studies did not reach the required offsets. Kaneda et al. (2010) found deeper reflectors at a depth of 35 to 45 km beneath seamounts with a hot spot origin in the northwestern Pacific Basin. They inferred that these reflectors indicate structures that formed via intraplate igneous activities, such as at the top of a mantle partial melting zone or a remnant of a hot spot plume head. OIB rock samples obtained from the Oki-Daito Ridge in the broad sense were different from the samples collected from the Amami Plateau and Daito Ridge (e.g., Ishizuka et al.2013). The existence of a deep reflector may therefore indicate intraplate volcanic activity in this region, although further investigation is required in order to confirm this interpretation.

The Oki-Daito Ridge shows a narrower ridge topography compared to the Oki-Daito Plateau and the Oki-Daito Rise; however, it has a thicker crust of 20 to 23 km (Figure 8a). The top of the middle crust, which has Vp of 6.3 to 6.8 km/s, is shallower beneath the northeastern bathymetric high. Slightly thinner crust and a notably higher Pn velocity were measured at the boundary between the ridge and the Minami-Daito Basin to the north along the KPr15 and DAr5 seismic lines. Apparently, a similar structure was found at the north of the Daito Ridge, where rifting is inferred. However, neither thin crust nor high Pn velocities were clearly observed at the transition from the Minami-Daito Basin along ODr5. The Pn velocity obtained beneath the Minami-Daito Basin varies between around 7.7 and 8.2 km/s and is locally higher than 8.0 km/s (Nishizawa et al. 2013). Therefore, the shallower and thicker middle crust beneath the northeastern side of the ridge may be related to the bathymetric high that was presumably produced by OIB activity 44 to 48 Ma ago (Ishizuka et al. 2013).

While the linear valley topography suggests rifting along the ridge, the velocity models do not show obvious axial symmetry. To confirm this feature, additional profiles across the ridge with symmetrical seafloor topography are needed.

Although large variations in crustal structure exist across the study area, though the crustal model for the Oki-Daito Rise constitutes an exception, we can summarize our results as follows. The Vp models obtained for the bathymetric highs in the Daito Ridges region generally correspond to a middle crust with Vp of 6.3 to 6.8 km/s, a lower crust with Vp of 6.8 to 7.2 km/s, a Pn velocity of 7.6 to 7.8 km/s, and a total crustal thickness of 15 to 25 km. Rock samples obtained from these Daito Ridges bathymetric highs indicate that most of them have originated from island arcs (e.g., Hickey-Vargas2005;Ishizuka and Yuasa 2007;Ishizuka et al. 2013).

There remains the question of whether our Vp models yield characteristics similar to the Japan island arc. Before 1990, the Japan island arc crust was considered to be composed of two layers, but it is currently divided into three parts owing to recent progress in observation and data analysis (Iwasaki and Sato 2009; right bottom panel in Figure 11). Iwasaki and Sato (2009) revealed that the crust of the Japan islands is composed of an upper part with a large Vp range of 5.5 to 6.1 km/s, a middle part with Vp of 6.2 to 6.5 km/s, and a lower part with Vp of 6.6 to 7.0 km/s. Furthermore, the Japan island arc crust has a lower crust $Vp < 7.0$ km/s, a low Pn velocity of 7.5 to 7.9 km/s, and a total crustal thickness of 30 to 40 km; thus, the middle and lower crusts have a significantly lower Vp than those of the Daito Ridges. Moreover, the Japan island arc has a much thicker crust, which indicates that it is a more mature island arc when compared to the Daito Ridges.

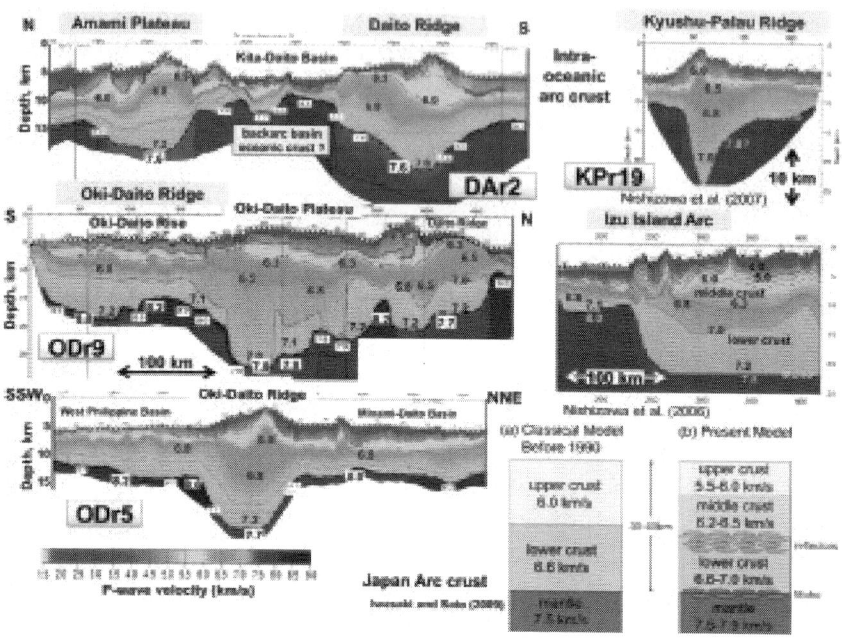

Figure 11: Comparison of Vp models for island arc crusts around Japan. Vp models in this study are shown on the left. Examples of intra-oceanic island arcs include the southern Kyushu-Palau Ridge (Nishizawa et al. 2007) (top right panel) and the Izu-Ogasawara island arc (Nishizawa et al. 2006) (middle right panel). Simplified models for Japan island arc are illustrated in the bottom right panel (modified from Iwasaki and Sato 2009).

In contrast, the Vp structure of the IBM island arc (an immature intra-oceanic island arc) is characterized by a middle crust with Vp of 6.0 to 6.8 km/s, a lower crust with $Vp \geq 7.0$ km/s at its base, and an uppermost mantle with $Vp < 8.0$ km/s (e.g., Suyehiro et al. 1996;Nishizawa et al.2006;Takahashi et al. 2007). We compared our Vp models for the Daito Ridges with those for several intra-oceanic island arcs near Japan (Figure 11). The Vp models of the Daito Ridges show a closer match to those of the IBM arc and Kyushu-Palau Ridge, which are immature paleo-island arcs, than they do to those of the mature Japan island arc. Moreover, the velocities of the middle and lower crusts are not as high as those of the Aleutian arc, which may be a result of extension and rifting in this region (Calvert 2011). Future studies on *S* wave velocity structure are expected to provide more information about the material properties of these paleo-island arcs.

AUTHORS' CONTRIBUTIONS

AN, KK, YK, and MO contributed to the analysis and interpretation of the seismic data. AN compiled the regional tectonic information and drafted the manuscript. All authors read and approved the final manuscript.

ACKNOWLEDGEMENTS

The authors gratefully acknowledge Prof. Emeritus J. Kasahara, Dr. R. Kubota, and the members of the Continental Shelf Surveys Co. Ltd. for the data processing and analysis. Discussions with Y. Ohara, K. Tani and T. Morishita were valuable. We thank the two anonymous reviewers for their insightful comments. The authors are also grateful to the members of the Hydrographic and Oceanographic Department, JCG for the management of the seismic surveys.

REFERENCES

1. Calvert AJ (2011) The seismic structure of island arc crust. In: Brown D, Ryan PD (eds) Arc-continent collision, Heidelberg: Frontiers in Earth Sciences. Springer. pp 87-119 10.1007/978-3-540-88558-0_4
2. Canales JP, Tucholke BE, Xu M, Collins JA, DuBois DL (2008) Seismic evidence for large-scale compositional heterogeneity of oceanic core complexes. Geochem Geophys Geosyst 9:1-22 doi:10.1029/2008GC00200
3. Deschamps A, Lallemand S (2002) The West Philippine Basin: an Eocene to early Oligocene back arc basin opened between two opposed subduction zones. J Geophys Res. doi:10.1029/2001JB001706
4. Fujie G, Kasahara J, Sato T, Mochizuki K (2000) Traveltime and raypath computation: a new method in a heterogeneous medium. J Soc Explor Geophys Jpn 53:1-11
5. Fujie G, Ito A, Kodaira S, Takahashi N, Kaneda Y (2006) Confirming sharp bending of the Pacific plate in the northern Japan trench

subduction zone by applying a traveltime mapping method. Phys Earth Planet Inter 157:72-85 doi:10.1016/j.pepi.2006.03.013

6. Hickey-Vargas R (2005) Basalt and tonalite from the Amami Plateau, northern West Philippine Basin: new Early Cretaceous ages and geochemical results, and their petrologic and tectonic implications. Island Arc 14:653-665
7. Ishihara T, Koda K (2007) Variation of crustal thickness in the Philippine Sea deduced from three-dimensional gravity modeling. Island Arc 16:322-337
8. Ishizuka O, Yuasa M (2007) Age and geochemical characteristics of the igneous rocks recovered the Philippine Sea and Ogasawara Plateau area as a part of basic researches on exploration technologies for deep-sea natural resources. JOGMEC, Tokyo: Rep on Deep Sea Survey Technology for Natural Resources in Japan. pp 255-296
9. Ishizuka O, Taylor RN, Ohara Y, Yuasa M (2013) Upwelling rifting and age-progressive magmatism from the Oki-Daito mantle plume. Geology 41(9):1011-1014 doi:10.1130/G34525.1
10. Iwasaki T, Sato H (2009) Crust and uppermantle structure of island arc being elucidated from seismic profiling with controlled sources in Japan. J Seism Soc Jpn (Zisin) 61:S165-S176 Japanese with English abstract
11. Kaneda K, Kodaira S, Nishizawa A, Morishita T, Takahashi N (2010) Structural evolution of preexisting oceanic crust through intraplate igneous activities in the Marcus Wake seamount chain. Geochem Geophys Geosyst 11: Q10014 doi:10.1029/2010GC003231
12. Kasuga S, Koyama K, Kaneko Y (1986) Geomagnetic and gravity anomalies around the Daito Ridge. Rept Hydrogrp Res 21:65-76 Japanese with English abstract
13. Kobayashi K, Kasuga S, Okino K (1995) Shikoku basin and its margins. In: Taylor B (ed) Backarc basins: tectonics and magmatism, New York: Plenum. pp 381-405
14. Korenaga J, Holbrook WS, Kent GM, Kelemen PB, Detrick RS, Larsen HC, Hopper JR, Dahl-Jensen T (2000) Crustal structure of the southeast Greenland margin from joint refraction and reflection seismic tomography. J Geophys Res 105:21591-21614 doi:10.1029/2000JB900188

15. Kubota R, Nishiyama E, Murase K, Kasahara J (2009) Traveltime estimation of first arrivals and later phases using the modified graph method for a crustal structure analysis. Explor Geophys 40:105-113
16. Larsen SC, Schultz CA (1995) ELAS3D: 2D/3D elastic finite-difference wave propagation code. Lawrence Livermore National Laboratory technical report no. UCRL-MA-121792. 19
17. Murauchi S, Den N, Asano S, Hotta H, Yoshii T, Asanuma T, Hagiwara K, Ichikawa K, Sato T, Ludwig WJ, Ewing JI, Edgar NT, Houtz RE (1968) Crustal structure of the Philippine Sea. J Geophys Res 73:3143-3171
18. Nishizawa A, Suyehiro K, Shimizu H (1983) Seismic refraction experiment at the Amami Plateau. J Phys Earth 31:159-171
19. Nishizawa A, Kaneda K, Nakanishi A, Takahashi N, Kodaira S (2006) Crustal structure of the ocean–island arc transition at the mid Izu-Ogasawara (Bonin) arc margin. Earth Planets Space 58:e33-e36
20. Nishizawa A, Kaneda K, Katagiri Y, Kasahara J (2007) Variation in crustal structure along the Kyushu-Palau Ridge at 15–21 N on the Philippine Sea plate based on seismic refraction profile. Earth Planets Space 59:e17-e20
21. Nishizawa A, Kaneda K, Oikawa M (2009) Seismic structure of the northern end of the Ryukyu Trench subduction zone, southeast of Kyushu, Japan. Earth Planets Space 61:e37-e40
22. Nishizawa A, Kaneda K, Oikawa M (2011) Backarc basin oceanic crust and uppermost mantle seismic velocity structure of the Shikoku Basin, south of Japan. Earth Planets Space 63:151-155
23. Nishizawa A, Kaneda K, Oikawa M (2012) T51D-2622: compilation of seismic structural models of the Kyushu-Palau Ridge, paleo-island arc in the Philippine Sea plate, at 13-30 N. In: AGU fall meeting abstract. San Francisco. 3–7 Dec 2012
24. Nishizawa A, Kaneda K, Oikawa M (2013) SCG67–08: seismic structure of the Kita-Daito Basin and Minami-Daito Basin in the northwestern Philippine Sea plate. In: JpGU meeting abstract. Chiba. 19–24 May, 2013
25. Okino K, Kato Y (1992) What is the Oki-Daito Ridge? Rept Hydrogrp Res 28:269-292 Japanese with English abstract

26. Okino Y, Ohara Y, Kasuga S, Kato Y (1999) The Philippine Sea: new survey results reveal the structure and the history of the marginal basins. Geophys Res Lett 26:2287-2290
27. Suyehiro K, Takahashi N, Ariie Y, Yokoi Y, Hino R, Shinohara M, Kanazawa T, Hirata N, Tokuyama H, Taira A (1996) Continental crust, crustal underplating, and low-Q upper mantle beneath an oceanic island arc. Science 272:390-392
28. Takahashi N, Kodaira S, Klemperer S, Tatsumi Y, Kaneda Y, Suyehiro K (2007) Crustal structure and evolution of the Mariana intra-oceanic island arc. Geology 35:203-206
29. Tokuyama H, Kagami H, Nasu N (1986) Marine geology and subcrustal structure of the Shikoku Basin and Daito Ridges Region in the northern Philippine Sea. Bull Ocean Res Inst Univ Tokyo 22:1-169
30. Tokuyama H (2007) Tectonic development and reconstruction of Philippine Sea Plate since late Cretaceous. JOGMEC, Tokyo: Report on Deep Sea Survey Technology for Natural Resources in Japan. pp 430-456
31. Ueda Y (2004) Paleomagnetism of seamounts in the West Philippine Sea as inferred from correlation analysis of magnetic anomalies. Earth Planets Space 56:967-977

Chapter 6

Vhub: a Knowledge Management System to Facilitate Online Collaborative Volcano Modeling and Research

Jose L Palma[1, 2], Leah Courtland[3, 5], Sylvain Charbonnier[3], Riccardo Tortini[4], and Greg A Valentine[2]

[1]Departamento de Ciencias de la Tierra, Universidad de Concepción, Victor Lamas 1290, Concepción, Chile

[2]Department of Geology and Center for Geohazards Studies, 411 Cooke Hall, University at Buffalo, Buffalo, NY 14260, USA

[3]Department of Geology, University of South Florida, 4202 E. Fowler Avenue, Tampa, FL 33620-8100, USA

[4]Geological and Mining Engineering and Sciences, 630 Dow Environmental Sciences, Michigan Technological University, 1400 Townsend Drive, Houghton, MI 49931, USA

[5]School of Earth & Environmental Sciences, Georgia Institute of Technology, 311 Ferst Drive, Atlanta, GA 30332-0340, USA

ABSTRACT

Knowledge of volcanic systems and the hazards they produce is rapidly advancing as internet resources become more readily accessible, new and more sensitive field techniques are developed, and ever greater amounts of data are collected. Such rapid advances drive the need for an online collaborative knowledge management system that enables the sharing of volcanological information, and modeling and analysis tools. Vhub (http://vhub.org) is a community cyberinfrastructure platform designed for collaboration in volcanology research, education, outreach, and discovery that complements existing volcano databases and other cyberinfrastructure projects. Vhub is unique in its functionality as a nucleus for the creation of collaborative groups focused on issues such as code development, field research, education, and hazard mitigation. In addition, Vhub serves as a clearinghouse and virtual platform for computational tools relating to volcanic processes and data analysis, as well as documentation to aid in the use and understanding of these tools. By providing a means for scientists to easily disseminate data, models, and ideas, Vhub aims to promote collaboration amongst scientists and to provide resources for science education while advancing the state of understanding of volcanoes and the hazards they produce. This paper introduces the Vhub cyberinfrastructure and provides an overview of select hub features and resources.

INTRODUCTION

Volcanology has been undergoing a fundamental transformation over the past 10–15 years brought about by the internet's facilitation of access to large volumes of information, greater community acceptance of computational modeling and statistical analysis, and increased attention to volcanic hazards. Despite these trends, the development of scientific research and applications that require numerical computation is hampered by the relative isolation of individual researchers and research groups, and by a gap between research-level development of computational models and their deployment at observatories with real-world hazard mitigation responsibilities. The emergence of innumerable web-accessible resources has many positive aspects, but sthese resources often have low visibility, insufficient documentation, and/or require computational resources or expertise that would-be end users do not possess. Each of these issues represents a barrier to the development and application or adoption of computational tools. There is an urgent need for a better approach to confront these issues that fall under the general topic of collaborative knowledge management. Such an approach should take advantage of the rapid growth of resources, provide new mechanisms for global collaboration in research, and disseminate resulting tools to researchers, observatory practitioners, educators and stakeholders, while reducing barriers associated with local infrastructure, and supporting continued growth of more focused resources that are locally managed at sites around the world.

In the 1990s, the nanoscience research community recognized similar barriers to progress within their own field and so began the development of a cyberinfrastructure "hub" to serve as a central online resource promoting connections amongst researchers and facilitating accessibility to models and information. The result of their efforts, nanoHUB (http://nanohub.org), has had profound effects on the rate of progress of nanoscience, making research tool development and use easily accessible, and enabling the use of advanced research tools for educational purposes. To support these efforts, software developers at Purdue University created the HUBzero® Platform for Scientific Collaboration (McLennan and Kennell 2010), the cyberinfrastructure that provides nanoHUB with all the functionality to develop and run online simulation tools, create private groups for collaboration, manage digital resources including videos, presentations and wiki-style content, communicate through a basic email-like messaging system, and many other capabilities.

The Vhub (short for "Volcanology hub") cyberinfrastructure builds upon the experience and success of nanoHub to advance collaborative knowledge development, storage, and transfer within the volcanological community. By January 2014 Vhub had logged more than 15,000 users and we expect a further increase within the next 5 years, following the trend exhibited at nanoHub (see Table 1 for a comparison of Vhub and nanoHub usage metrics; note that Vhub was created 8 years after nanoHUB, and compare the number of users between nanHUB 2005 and Vhub 2013). At its core, Vhub is a community-driven platform whose content is created and managed by its users (e.g. Elmagarmid et al. 2008). As a heterogeneous science, volcanology encompasses a wide range of workers varying from those who are highly theoretical and computationally focused, to experimentalists, and to those whose work focuses on observational and analytical data (i.e., field geology, geophysics, geochemistry, and remote sensing). As a result, adoption of a hub environment by this community is complex and requires significant cross-disciplinary communication. Here we briefly describe and provide examples of some of the capabilities of Vhub that facilitate the advancement of basic and applied research currently being conducted by the diversity of work related to volcanoes, volcanic processes, and associated hazards.

Table 1: Comparison of usage statistics for nanoHUB during 2005 and 2013, and Vhub for 2013

	nanoHUB 2005	**nanoHUB 2013**	**Vhub 2013**
Registered users(1)	2,179	27,726	1,157
Unregistered interactive users(2)	5,380	34,088	1,632
Unregistered download users(3)	11,309	243,425	12,368
Total users(4)	18,868	304,639	15,157
Visitors(5)	50,926	593,332	42,669
Simulation users(6)	1,660	13,106	550
Simulation runs (jobs)	53,660	501,469	22,544

[1]Number of Registered Users that logged in. User registration assigns a unique login to each individual user.

[2]Number of Unregistered Users, identified by unique hosts/IPs, that had an active session (active on the site for at least 15 minutes) without logging in. Does not include known web bots/crawlers.

[3]Number of Unregistered users, identified by unique hosts/IPs, that had an active session of less than 15 minutes without logging in and downloaded a non-interactive resource such as PDF or podcast. Does not include web bots/crawlers.

[4]Sum of Registered Users, Unregistered Interactive Users and Unregistered Download Users.

[5]Identified by a unique IP address / hostname.

[6]Number of Registered Users[(1)] that ran one or more simulation runs.

Palma et al.

Palma et al. Journal of Applied Volcanology 2014 3:2, doi:10.1186/2191-5040-3-2

VHUB DEVELOPMENT AND VISION

Vhub was conceived in 2008 with an initial focus on increasing accessibility both to quantitative models of volcanic processes and hazards, and to data sets collected for volcanoes around the world. Specifically, a means was sought to facilitate knowledge and data transfer between volcanic observatories and other researchers. While this core mission remains, Vhub has grown to provide a broader platform for collaborative research and education in volcanology (Figure 1). The project was launched in 2010 with funding from the US National Science Foundation and includes a core development team at University at Buffalo, Michigan Technological University, and University of South Florida, along with a group of collaborators from the international community (https://vhub.org/about/vhubteam). The basic software infrastructure is maintained at Purdue University where the primary Vhub servers reside (see https://hubzero.org).

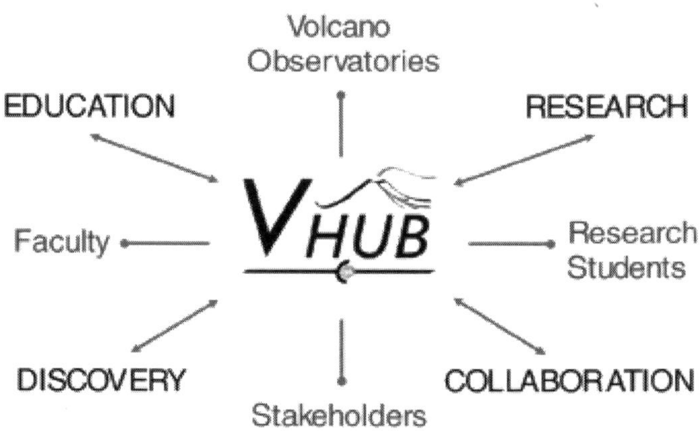

Figure 1: Vhub incorporates a variety of features designed to connect faculty at research institutions with students, scientists at volcano observatories with

stakeholders, and faculty at research institutions with scientists at observatories and other institutions in charge of hazard and/or risk mitigation.

Vhub provides infrastructure for knowledge management, defined as "the creation, archiving and sharing of valued information, expertise and insight within and across communities of people and organizations with similar interests and needs" (Rosenberg 2001). In this context, Vhub development efforts have been consistent with the four general objectives of knowledge management projects (Davenport et al. 1998): 1) create knowledge repositories, 2) improve knowledge access, 3) establish an environment conductive to more effective knowledge creation, transfer and use, and 4) manage knowledge as an asset. Note that we use the word *knowledge* to mean not just information (data), but also tools to improve understanding and application of that understanding. Vhub is designed to function as a platform for sharing a variety of forms of knowledge, including a unique emphasis on computational tools pertaining to volcanology, and as a platform that facilitates collaboration. Vhub itself is not a database; databases can be shared on the platform, but Vhub aims principally to provide capabilities that complement rather than duplicate existing efforts centered specifically on compiling and managing data about volcanoes, such as the Global Volcanism Program (http://www.volcano.si.edu; Siebert et al. 2010), the Volcano Global Risk Identification & Analysis Project (http://www.bgs.ac.uk/vogripa/; Crosweller et al. 2012). Instead, Vhub is a portal to simulations, lectures, existing databases, educational resources, and other content of interest.

CYBERINFRASTRUCTURE AND MEMBERSHIP

Vhub takes advantage of the HUBzero® open-source software platform. This software incorporates several well-known open-source packages, including an Apache web server, the Joomla content management system, PHP web scripting, and a MySQL database for storing content and usage statistics (Klimeck et al. 2008). The software is specifically designed to support scientific collaboration, research, education, and discovery, the four pillars of Vhub's organization (Figure 1). HUBzero can handle a variety of data types including documents created with

Microsoft Office©, PDFs, ASCII files, images and video clips, analysis tools, and datasets. Additionally, all uploaded content is integrated with collaboration features such as forums, wikis, blogs and Q&A. Integrating all of these packages, the Vhub web site serves as the interface between the user and the platform.

While many Vhub resources are available to the public for viewing or downloading (Table 2), membership is necessary to take full advantage of the platform. Becoming a member of the Vhub community is easy. The registration process is automatic and requires a minimum amount of personal information: login name, password, e-mail and employment status. Other information such as website url, telephone number, institution, and interests is optional, although useful for tracking Vhub's use. Once registered, members are assigned an ID number that can be used to access their profile, contributions, usage statistics, and personal blog. For example, the ID number of one of the authors of this article is 1097, and the url to his page is https://vhub.org/members/1097; to access his personal blog simply add "/blog" at the end of that url. Each Vhub member has a profile useful for sharing contact details, interests, and a short biography. All members are also provided with an internal messaging system to strengthen communication and collaboration between members. This system works like a basic web-based email client, and includes Inbox, Sent, Archive, and Trash folders. The user can set up which messages go to Vhub's internal Inbox and which are forwarded to an external email address. When composing a message the user can easily select any number of recipients from the list of Vhub members.

Table 2: Description of the different resource types available within Vhub

Resource type(1)	**Description**
Courses	A collection of lectures and/or other types of educational materials that make a course on a specific topic.
Data sets/ Collections	Repository of data from volcanic experiments, modeling, field trips, GIS, hazard mapping or other related work.

Educational materials	Individual educational materials aimed at teaching a specific subject such as lectures and homework assignments. For example, it can be a narrated presentation or a written document. They may also include a short video or web link to other type of media. Generally they are part of a set of complementary resources on a specific topic.
Online tools	A software application that allows users to run a specific type of calculation directly on VHub resources within your browser. They include numerical simulations of volcanic flows, statistical tools and algorithms to process and plot petrological data.
Offline tools	A downloadable software application that users can run on their own computer, subject to platform/software requirements.
Presentations	A lecture, seminar or other type of presentation in various types of media. Useful for sharing material presented at conferences or workshops.
Publications	A document relevant to the community that has been published in some manner. It includes white papers, theses, pre-prints, and magazine articles.
Workshops	A collection of lectures, seminars, and materials that were presented at a workshop.
Miscellaneous	Any resource that might not easily fit into any of the previous categories.
Topic pages and articles[(2)]	Wiki articles for explaining terminology, methodologies, resources or any topic of interest to the community.
Vhub blog[s(3)]	A public blog that any member can contribute to. It may be used, for example, to report volcanic activity or to describe meetings, workshops or courses. Members can upload their comments.
Personal blogs[(4)]	Every member can manage a personal blog in which only the member can add entries. They can be accessed by anyone.

[1]The first nine categories can be accessed at https://vhub.org/resources.

[2]Accessible at https://vhub.org/explore/wiki.

[3]Accessible at https://vhub.org/explore/blogs.

[4]Located in the Blog tab of the My Account page of every user. For example: https://vhub.org/members/1097/blog

Palma et al.

Palma et al. Journal of Applied Volcanology 2014 3:2, doi:10.1186/2191-5040-3-2

A personalizable user panel, or 'dashboard', is included to allow users to organize resources and user-specific information. Connecting via this interface, users view and interact with the site without needing to be aware of the complex system of software, data-storage devices, supercomputing hardware and other middleware that host and manage the content being accessed. Vhub members can thus focus only on the resources of interest to them while working towards their individual research, monitoring, or hazard mitigation goals. The dashboard can be accessed through the "My Account" link located at the top of the page. In addition to accessing a customizable dashboard and communication resources, establishing a membership account with Vhub will automatically allot users 1 Gb of disk space on hub servers, enabling them to run online simulation tools via the hub infrastructure.

GROUPS, COLLABORATION, AND SHARING

Vhub provides capabilities for online collaboration in a variety of ways, including (1) within teams conducting basic research, (2) between participants and external advisors on complex hazard assessment projects, (3) between mentors/teachers and students, and (4) amongst community stakeholders. All such collaborations can have personalized privacy levels.

Collaboration can be restricted to a small audience through the creation of 'groups'. Each group has a dedicated url, a name, and a logo that are set when the group is created. Any registered user can create a group, set up the content of the group, and specify who can join. The user that creates the group automatically becomes a group manager. Managers can change the description and content of group pages, send messages to all or select members, cancel memberships, promote members to managers, and assign member roles.

Group privacy settings are controlled by group manager(s), including the option to determine what content is visible to the general public, to Vhub account holders, and to group members. If designated 'visible', groups can be found in searches as well as by browsing (https://vhub.org/groups/browse). If set to 'hidden', the group is only discoverable by group members. Unique access restrictions can be set to group content as well.

Four options are available to restrict the access to a group: "Anyone", "Restricted", "Invite Only" and "Closed". This allows options for membership to be requested by users and approved automatically, requested by users and approved by a group manager, initiated by group managers and accepted by users with no opportunity for users to request access, or closed such that no new members are accepted to the group.

Groups may have different purposes. Organizations such as the Commission on Monogenetic Volcanism of the International Association of Volcanology and Chemistry of the Earth's Interior, IAVCEI (https://vhub.org/groups/iavcei_cmv) and the Latin-American Volcanological Association, ALVO (https://vhub.org/groups/alvo) manage groups that operate as portals to the information and activities that they promote. Research teams may set up groups to discuss their progress and internally disseminate documents and ideas. Educators may decide to manage volcanology or hazards courses and include lectures, homework assignments, and class discussions as content. Groups can also be set up to restrict access to resources such as presentations and online computational tools, thereby keeping data and sensitive information private. Vhub groups have a variety of features to facilitate their work (Table 3).

Table 3: Description of the features available for Vhub groups

Group feature	**Description**
Overview	This section provides a general description of the group. It may also inform about membership restrictions and have announcements to the members. Subpages within this section may be created, for example to have a version of the overview in a different language, or to organize specific information about the group. The overview can only be edited by the managers of the group.

Members	List of all the members of the group. It includes the status (whether the member is a managers of the group or not) and the role of the member if it has been assigned. Managers can send messages to one or all members.
Wiki	Wiki pages are single pages that any member can create and comment on. It supports wiki formatting and allows the insertion of figures and documents. The content of these pages depends on the scope and organization of the group.
Resources	Presentations, data, documents, etc., that are uploaded by members of the group. When the resource is uploaded the author must select the group that the resource belongs to in the Group ownership section of the uploading process. Since all resources on Vhub have a unique *url*, members can add a direct link to the resource in any of the other group features.
Discussion	In this section a member can start or follow up on a discussion in any topic relevant to the group. Once a discussion has been started, other members can add comments, questions, answers, or further information on the topic. The managers can create categories to maintain the discussing organized. In other websites this capability is known as Forum.
Messages	This section contains all messages sent by the managers of the group to all group members.
Blog	This feature allows members to add communications of interest to other members of the group. Other members can add comments to the original posts. Similar to other blog services on the web, the posts are organized by date and the most recent and popular blogs are highlighted.
Wish list	A group wish is generally a suggestion for improvement that focuses on the content and work within the group. Any member can send a wish, comment and like or dislike existing wishes. A wish can then be accepted, rejected or granted.
Calendar	To add and discover upcoming events or activities related to collaboration and communication within the group.

Palma et al.

Palma et al. Journal of Applied Volcanology 2014 3:2, doi:10.1186/2191-5040-3-2

Vhub's Wiki pages and Blogs are additional means of communication that members of the platform can take advantage of (Table 2). While Vhub blogs are public and designed to contain information relevant to the general community, individual members are also provided dedicated blog space for personal communication; privacy settings for individual posts can be set when they are created or edited. Blogs can be used, for example, to report descriptions of volcanic activity or field trips, or to share personal points of view on specific subjects. A Wiki is a web-based hypertext system which supports community-oriented authoring that allows collaborative content creation (Shih et al. 2008). Vhub Wiki Pages and Articles may receive contributions from interested members either through direct editing of the wiki or through suggestions. Topic pages can be used, for example, as tutorial articles describing how laboratory techniques or computational tools work, to discuss the definition and proper use of volcanological terms, or to describe and discuss volcanological concepts (an example focusing on volcano remote sensing capabilities ishttps://vhub.org/topics/SatelliteMonitoring).

KNOWLEDGE REPOSITORY AND SUPPORT NETWORK

Vhub hosts a Resource Warehouse organized into nine categories based on the type of content submitted: Courses, Data Sets/Collections, Education Materials, Offline Tools, Online Tools, Presentations, Publications, Workshops, and Miscellaneous. See Table 2 for a short description of each resource type. These contents are copyrighted by their authors and contributors, who give Vhub license to display and distribute the material according to the agreement detailed on the following webpage: https://vhub.org/legal/license *webcite*. All resources on Vhub include a title, author list, abstract, and archival citation information. They are assigned a dedicated url that can be used for citation in journal publications (e.g. Rose 2011). Files uploaded to any individual resource are also assigned an exclusive url, which can be linked to from its abstract or from other resources. While there are no restrictions on the type of files that may be uploaded to the hub, in order to promote accessibility contributors are encouraged to avoid uploading files in proprietary formats that require commercial software for viewing. For example, it is recommended to save presentations and Microsoft Word documents as PDF documents before uploading to vhub.org. As part of the process of creating a resource, the contributor selects 'tags' to categorize the content. Tags are simply key words assigned to a resource by the user making the contribution. Tags are searchable and thus aid in resource discoverability.

Each time a new resource is created, the contributor must set the access level of its content. 'Public' resources grant full access, including access to the abstract and to any supporting documents uploaded with the resource, to anyone visiting the web page of that resource. When the resource is set to 'protected', the title and author list are visible to everyone but the abstract, references, citation information and any associated documents are only accessible by members of the specific group with which the resource is aligned. Like protected resources, 'private' resources must be aligned with a group. However in the case of private resources, only group members can discover and access the resource; in this case the resource and all associated content is hidden from all other users. Through these classifications, contributors have full control over whether a given resource is made available to the public or only to specified collaborators.

Since Vhub is a community-driven knowledge environment, we expect the community to contribute to this knowledge system and to learn from it at the same time. This increases the impact of collaboration and opens up new possibilities for users with few resources to gain access to collaboration pathways and shared resources.

ONLINE SIMULATION TOOLS

The concept of knowledge transfer within Vhub includes development and dissemination of tools that promote the analysis and understanding of volcanic processes. To that end, one important feature of Vhub is its capacity to execute simulation, modeling, and data analysis programs within the user's web browser while the program being executed runs on remote computing resources, negating the need for the user to install software on their local machine. These 'Online simulation tools' include a graphical user interface (GUI) and documentation to facilitate and encourage volcanologists, particularly those who do not regularly use modeling tools, not only to take advantage of the computational resources available to them but to use each tool appropriately. The development and use of computational tools can benefit from the many collaboration features available on Vhub including discussion panels, internal messaging, wish lists, wiki pages, and Q&A sections. These tools help to bring together program developers, experts in the field, tool users, and educators. To facilitate appropriate use of modeling tools, Vhub encourages developers and experts to describe what the models can and cannot do, their advantages over other techniques, and their limitations, as well as to provide advice on how to read and use the model results.

Questions asked from within an online tool are linked to that tool and can be accessed through the "Questions" tab on the resource page (e.g. https://vhub.org/resources/pingu/questions). The developers of the tool get an email every time a question about their tool is posted. Unless the question is closed, any member can answer a question and contribute to solve problems using a tool. Finding or posting questions (and their answers) for specific tools is easy for regular users and developers, and facilitates feedback and collaboration. All questions are listed under "Questions and Answers" in the Explore submenu.

In order to facilitate implementation of numerical simulations and/or processing algorithms by users, Vhub hosts a library of educational and training material prepared by tool developers and expert users. This material may exist in video format, as a topic wiki page, or as a downloadable pdf or other type document. Course and/or workshop materials also provide background information on the theory and practical applications of simulation tools. The developers of the tools, as well as the authors of any other resource material, can track the number of users that have accessed their contribution and obtain feedback from the community.

Additionally, tool users have the ability to ask Vhub community members to view and comment on simulation results. Researchers and students can participate in discussions and establish new collaboration paths or suggest improvements to existing tools. Findings can be shared with the community within blog or topic pages. Educators can use resources to facilitate student learning in a classroom environment (e.g. Courtland et al. 2012). From interactions like these, innumerable collaborative research opportunities emerge, including opportunities for improving both the work of those who contribute to Vhub's knowledge repository and the users that benefit from that work.

Some of the computational tools currently available to Vhub users include: Energy Cone (rapid delineation of the impact zone by pyroclastic density currents; Sheridan and Macías 1995 Palma2013a), Tephra2 (tephra dispersion forecast tool; Bonadonna et al. 2005 Bonadonna et al. 2013), Bent (atmospheric plume analysis; Bursik 2001;

Bursik 2013), Hazmap (tephra dispersion forecast tool; Macedonio et al. 2005; Costa 2013), Titan2D (mass flow simulation tool; Pitman et al. 2003), PINGU (tool for plotting geochemical and petrological data; Cortés and Palma 2013) and BET_VH (Bayesian Event Tree analysis for Volcanic Hazards; Marzocchi et al. 2010; Tonini et al. 2012). The list of online simulation tools available on Vhub is expected to expand considerably as more volcanologists contribute tools and in so doing realize much more rapid dissemination of their work than through traditional routes (such as publishing a paper about a model and then providing the software upon request). Vhub is also becoming a central library for offline tools, which in some cases can be downloaded directly from the Vhub site (e.g., the Plumeria eruption column model; Mastin2011) or in others provide links to the author(s)'s website where source code is available for download.

Some online simulation tools create files that are stored in the user's account within Vhub, and they are kept after the user ends the tool session. These files can be accessed and managed, at any time, through the Workspace tool or through WebDav (see https://vhub.org/kb/tips/webdav). Below and in the Appendix we illustrate usage of the online tools Workspace, Tephra2, and Titan2D, and the offline tool OMIplot.

The Workspace and Model Development

The Workspace is a Linux desktop environment that can be used to upload and download files stored in the user's account, usually created by other online simulation tools (files can also be accessed through webdav, see https://vhub.org/kb/tips/webdav). The Workspace is also the perfect environment to edit, compile and test a tool in development. Developing online simulation tools within Vhub is achieved through a content-management-system for tool publication which provides a project area with wikis, a timeline that documents changes to each revision, and a subversion source-code repository (McLennan and Kennel 2010). Developers can chose whether they want to make source code private or accessible to other Vhub members, and can specify an open-source or other license. Once a tool is ready, it is committed by the developer and installed by Vhub's server managers, at which point the developers are asked to test the working version of the tool before it is fully deployed on the hub. More information on how to create a tool for Vhub can be found athttps://vhub.org/support/knowledgebase/development.

Modeling Tephra Fallout with Tephra2

Tephra2 is an advection–diffusion model used to forecast the accumulation of ejected volcanic material on the ground (Bonadonna et al. 2005). It can be used for civil protection purposes, such as giving public warnings and planning mitigation measures (Connor et al. 2001; Volentik et al. 2009), as a research tool for studying eruptions past and present (Johnston et al. 2012; Biass and Bonadonna 2011; Longchamp et al. 2011; Connor and Connor 2006; Volentik et al. 2010), or as an educational tool for volcanology, numerical modeling, or quantitative literacy (Courtland et al. 2012).

Users can access the Tephra2 code via either of two GUIs: 'Tephra2', which grants full access to the model and is intended primarily for hazards scientists and graduate students (Figure 2), and 'Tephra2: Student Version' which places limitations on code input and is intended primarily for high school through undergraduate students (Courtland et al. 2012).

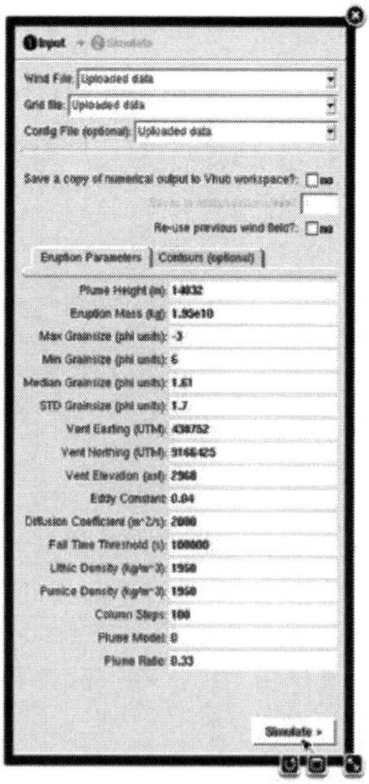

Figure 2: Graphical user Interface (GUI) for Tephra2. Values shown are used to calculate the Merapi example outlined in the Appendix. The wind file and

grid file must be uploaded from the user's computer. The configuration file may either be uploaded or the values of each parameter may be typed directly into the GUI. Output may be saved to storage on Vhub with the option 'Save a copy of numerical output to Vhub workspace?' selected. Contours to plot are listed under the second tab. Only select 're-use previous wind field?' when you wish to re-apply one of the 'random' wind fields pre-loaded into the GUI. Default is to select a new wind field each time the model is run with the 'random' option selected.

The Tephra2 online simulation tool applies equations for particle diffusion, transport, and sedimentation to a set of user-supplied eruption parameters. After ejection from the vent, particles undergo column diffusion before being released into an atmosphere composed of a series of vertically stacked layers within which wind speed and direction are constant but between which they are allowed to vary. While falling, particles experience diffusion due to atmospheric turbulence and advection due to the wind before falling onto the land surface.

Atmospheric data required by the model can either be downloaded from the NCEP-DEO Reanalysis website directly (http://www.esrl.noaa.gov; Kanamitsu et al. 2002), obtained and formatted for use by the Windre online simulation tool (https://vhub.org/resources/windre), or typed into a text file that includes altitude (m a.s.l.), direction (0–360 degrees from north) and magnitude (m s^{-1}) of the local wind field. The easting, northing, and elevation of the ground surface at which to make calculations are listed in a separate 'grid' file. The remaining parameters (e.g., maximum height of eruption column, total mass of tephra ejected, pyroclast density, size distribution of particles) can either be defined in a third text file or typed into the appropriate GUI field (Figure 2). Each file is uploaded to the Tephra2 graphical user interface. Execution time scales with the number of points in the grid file. Example atmospheric profiles, grids, and configuration files for three volcanoes are provided with the tool. Users may either alter the input parameters for these example eruptions or provide information for the eruptive scenario of their choice.

Graphical output is in the form of an isomass map (Figure 3), which allows the user to quickly visualize results of a calculation. The numerical output on which this map is based is also available for download. An example, calculated for Merapi volcano, Indonesia, is provided in the Appendix.

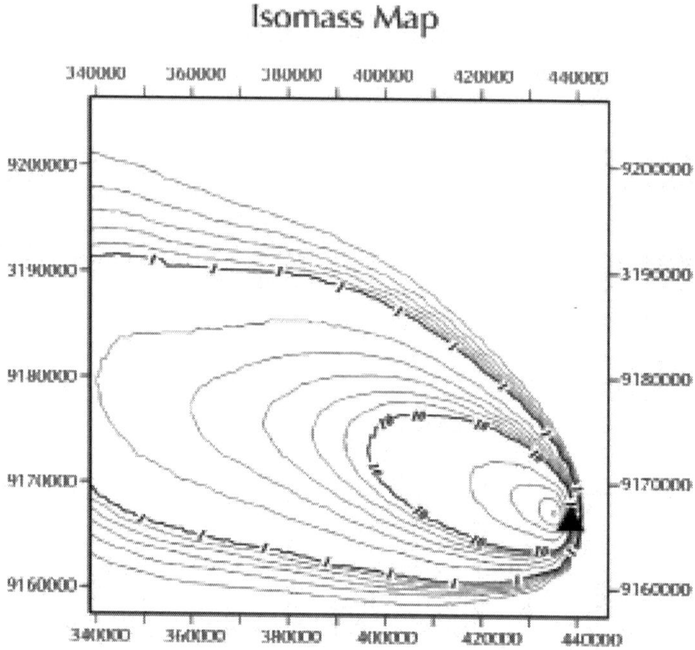

Wind input

06	2.0	339.8
774	2.7	322.3
1506	2.7	303.9
3145	5.1	280.5
4413	5.0	281.7
5870	6.1	292.8
7598	8.0	284.8
9714	9.1	281.7
10988	9.6	271.6
12476	10.8	260.8
14209	13.0	268.7
16613	14.6	297.0
18659	2.1	262.2
20673	1.6	77.4
23855	7.3	100.9
26455	9.3	116.5
31068	17.5	82.3

Figure 3: Wind input to and graphical output from the Tephra2 model. Wind input taken from renalysis data for Nov 4, 2010. Contours represent kg m^{-2} tephra accumulation on the ground. Easting coordinates are displayed on the x axis, northing coordinates on the y axis. Isomass units are kg m^{-2}. The map is automatically truncated to show only the locations where a user-defined minimum of tephra has accumulated, and thus may represent significantly less area than supplied with the input grid. Numerical output (not included, but available through the GUI) provides values for tephra accumulation at each grid point.

Simulating Dry Granular Flows over Natural Terrain with TITAN2D

Titan2D simulates dry granular avalanches over digital representations of natural terrain (Pitman et al. 2003; Patra et al. 2005). The program is based upon a depth-averaged model for an incompressible Coulomb continuum, a "shallow water" granular flow, based on the work of Savage and Hutter (1989), Iverson (1997), Iverson and Denlinger (2001), Denlinger and Iverson (2001) and Mangeney-Castelnau et al. (2002). This assumption is grounded in the fact that the flow thickness is small compared to the entire area over which a long-run-out flow travels and deposits material. A principal feature of the code is the incorporation of topographical data into the simulations and grid structure, combining numerical flow simulations with digital elevation data of natural terrain. A pre-processing routine allows digital elevation data to be imported. These data define a two-dimensional spatial box in which the simulation will occur. The raw data provide elevations at specified locations. Using these data, interpolating between data points where necessary, a rectangular Cartesian mesh is generated. Adaptive gridding allows for the concentration of computing power on a region of interest. Mesh refinement captures complex flow features at its leading edge, as well as locations where the topography changes rapidly. The conservation equations for mass and momentum are solved with a Coulomb-type friction term for the interactions between grains and between the granular material and the basal surface. Finally, the grid data are used, together with simulation output, in post-computation visualization (Yu et al. 2009; Charbonnier and Gertisser 2011 2012).

The Titan2D online simulation tool on Vhub can execute a complete simulation but it does not prepare the necessary digital elevation model (DEM) that Titan2D relies upon. Vhub offers an online conversion tool that puts a DEM into a format that the model can read and use (http://vhub.org/resources/dem4titan). A tutorial (http://vhub.org/resources/761*webcite*) describes how to enter input parameters for Titan2D simulations on Vhub to run a simulation, and to visualize the results using the Titan2D viewer (https://vhub.org/tools/titanview). Within the Vhub environment, the Titan2D numerical algorithm, its GUI, and supporting software packages are all installed. The Titan2D GUI on Vhub (Figure 4) allows the user to provide data that determine the flow behavior. Code execution can occur either within the designated Vhub server hosted at Purdue (local execution in the GUI job submission tab), or on a large cluster hosted at the University at Buffalo and transparently linked to the Vhub architecture (the Hub-submit option in the GUI).

Figure 4: Screenshot of the Titan2d Java-based graphical user interface (GUI) available within the online Vhub environment. From this GUI, the user can: (1) enter and save the different Titan2d input parameters, (2) run a Titan2d simulation and (3) monitor the progress of a simulation.

Output of the model can be visualized using the Titan2D viewer, which produces a map of flow depths overlaid on topography. One of the metrics used to evaluate the output of numerical simulations of this type is to calculate the 'flow coverage match', i.e. the correlation between actual and numerical flow coverage areas. The flow coverage match is defined by the region of mutual overlap (the intersection) divided by the total area of the combined numerical and field-mapped flow events (i.e. their union). Results obtained for one of the June 2006 pyroclastic flows at Merapi volcano show a flow coverage match of 54.4% (Figure 5). The details of this example are provided in the Appendix.

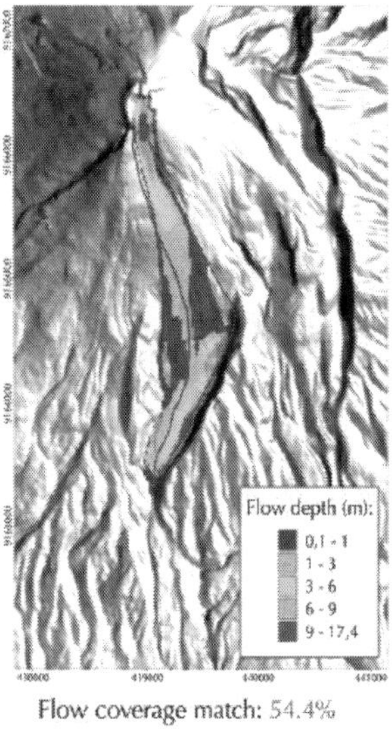

Figure 5: Titan2D inundation map obtained from the simulation of a short-to-medium-runout block-and-ash flow (SM-BAF) at Merapi Volcano showing the distribution of the maximum flow depths after 120 seconds (see text for explanation). The red outline is the mapped extent of the lobe 5 BAF deposits emplaced in June 2006 (Charbonnier and Gertisser, 2012). The 'flow coverage match' is obtained by dividing the intersection area of the mapped and simulated inundation areas by their union area.

Remote Sensing Of So_2 Emissions with Omi Data and OMiplot

OMIplot (Carn 2011) is a software package designed to ingest, visualize and analyze SO_2 (i.e. sulfur dioxide) and other trace gas data collected with OMI: a hyperspectral UV and visible spectrometer aboard NASA's Aura spacecraft. Aura is part of the A-Train satellite constellation and operates in a sun-synchronous, near polar orbit at an altitude of ~705 km (Levelt et al. 2006a). Each OMI orbit has a swath width of 2600 km and pixel size of 13 × 24 km at nadir (Levelt et al. 2006b), allowing continuous global measurements of ozone and various other trace gases including SO_2, nitrogen dioxide (NO_2) and bromine monoxide (BrO) (Carn et al. 2007). In clear sky conditions, OMI can provide daily remote measurements of volcanic SO_2 emissions if they exceed the sensor's detection limit (Carn et al. 2008). Level 2 OMI SO_2 (OMSO2) data files in HDF-EOS5 format are publicly available at no cost and downloadable from the NASA Goddard Earth Sciences Data and Information Services Center website (http://disc.sci.gsfc.nasa.gov/Aura/data-holdings/OMI/omso2_v003.shtml).

The functionality of OMIplot is specifically targeted to volcanologists, primarily for monitoring volcanic degassing and drifting eruption clouds. In particular, OMIplot can produce images of the SO_2 vertical column density (VCD) and other measurements e.g. Aerosol Index over a user-defined geographic region and calculate the SO_2 mass in volcanic plumes by integrating the SO_2 VCDs over the plume area (Lopez et al. 2012). The software can also be used to generate temporal averages and summations of OMI SO_2 measurements to improve signal-to-noise ratios, and to produce KMZ files for use with Google Earth. OMI SO_2 algorithms are subject to continuous refinement and OMIplot is in a continual state of development. Support for additional OMI data products (including NO_2, BrO, and aerosol) is being added to OMIplot, and a GNU Data Language (GDL) version is also forthcoming to permit online processing of OMI data in Vhub. Currently OMIplot is an offline tool on Vhub (Carn2011) available for download and use on local machines, provided Interactive Data Language (IDL) version 6 or later is installed. Outputs of OMIplot show SO_2 loading at different altitudes (Figure 6), as described in the Appendix.

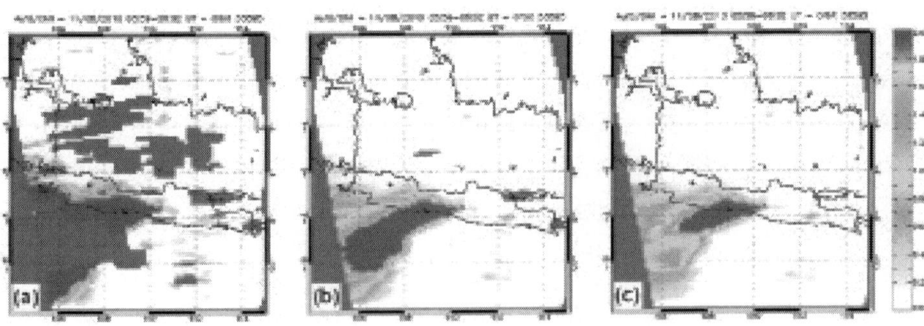

Figure 6: Plots of SO2 VCD over Java, Indonesia on November 8, 2010 generated by OMIplot. Merapi volcano (central Java) is indicated by a triangle. (a) Lower troposphere (TRL) SO_2 data; (b) Mid-troposphere (TRM) SO_2 data; (c) Upper troposphere/lower stratosphere (STL) SO_2 data. Gray regions denote data gaps, mostly due to meteorological clouds masking SO_2 at lower

assumed altitudes, and are more apparent in the TRL data due to the lower a-priori SO_2 center of mass altitude. The OMI row anomaly can also result in data gaps (and hence SO_2 mass underestimation) depending on the date and viewing geometry.

SUMMARY AND CONCLUSIONS

This paper provides a brief overview of many of the capabilities of Vhub, such as collaboration within groups, the deployment of modeling tools, and examples of online and offline tools. Vhub is a knowledge-management-system aimed at facilitating collaborative efforts to better understand volcanoes and to more closely link observatory practitioners with external resources as they emerge from the research community, and non-observatory researchers with the monitoring data and knowledge generated within volcano observatories. Vhub provides a mechanism for volcanologists to share knowledge and tools, and to rapidly connect users with the resources they need, whether they be students, academics, or volcano observatory personnel.

At its core, Vhub is a community driven effort that is constantly evolving in terms of the resources available within the hub, the ways that those resources are used, and the underlying cyberinfrastructure (software and hardware) itself. For example, significant current effort is focused on re-engineering core modeling tools for mass flows and tephra dispersion in order to make them more flexible for changing concepts and user needs, and in implementation of probabilistic models. Vhub goes beyond the traditional, static web presentation to provide an interactive platform capable of promoting explosive developments in the field of volcanology and volcano risk mitigation.

AUTHORS' CONTRIBUTIONS

JLP developed tools and wrote the sections on general aspects of Vhub. LC implemented Tephra2 and related tools and documentation on vhub and wrote the corresponding sections of the paper. SC contributed sections on Titan2D. RT contributed sections related to remote sensing tools. GAV coordinated the Vhub project and the compilation of the manuscript, and co-wrote introductory sections with JLP. All authors read and approved the final manuscript.

ACKNOWLEDGEMENTS

Vhub is supported by the U.S. National Science Foundation, through NSF Grant DRL 0940831. The Vhub development team includes, in addition to the authors: M.I. Bursik, E.S. Calder, S.M. Gallo, P. Johnson, M. Jones, J.V. Bajo, S. Melander, C. Renschler, A. Patra, J. Cortés, D. Moore-Russo, B. Pitman (University at Buffalo); C. Connor, L. Connor (University of South Florida); S. Carn, and W. Rose (Michigan Technological University). Numerous others in the volcanology community have contributed to the resources at Vhub.

REFERENCES

1. Biass S, Bonadonna C (2011) A quantitative uncertainty assessment of eruptive parameters derived from tephra deposits: the example of two large eruptions of Cotopaxi volcano, Ecuador. Bull Volcanol 73:73-90 doi:10.1007/s00445-010-0404-5
2. Bonadonna C, Connor CB, Houghton BF, Connor LJ, Byrne M, Laing A, Hincks T (2005) Probabilistic modeling of tephra dispersion: hazard assessment of a multiphase rhyolitic eruption at Tarawera, New Zealand. J Geophys Res 110:B03203 http://dx.doi.org/10.1029/2003JB002896
3. Bonadonna C, Connor LJ, Connor CB, Courtland LM (2013) Tephra2. https://vhub.org/resources/tephra2 *webcite*. Accessed 20 May 2013
4. Bursik MI (2001) Effect of winds on the rise height of volcanic plumes. Geo Res Lett 28(18):3621-3624
5. Bursik MI (2013) Bent - Atmospheric Plume Analysis. https://vhub.org/resources/bent Accessed 20 May 2013
6. Carn SA (2011) OMIplot Software. http://vhub.org/resources/682 *webcite*. Accessed 16 Apr 2013
7. Carn SA, Krotkov NA, Yang K, Hoff RM, Prata AJ, Krueger AJ, Loughlin SC, Levelt PF (2007) Extended observations of volcanic SO_2 and sulfate aerosol in the stratosphere. Atmos Chem Phys Discus 7:2857-2871

8. Carn SA, Krueger AJ, Krotkov NA, Arellano S, Yang K (2008) Daily monitoring of Ecuadorian volcanic degassing from space. J Volcanol Geotherm Res 176:141-150
9. Charbonnier SJ, Gertisser R (2009) Numerical simulations of block-and-ash flows using the TITAN2D flow model: examples from the 2006 eruption of Merapi Volcano, Java, Indonesia. Bull Volcanol 71:953-959
10. Charbonnier S, Gertisser R (2011) Deposit architecture and dynamics of the 2006 block-and-ash flows of Merapi Volcano, Java, Indonesia. Sedimentology 58:1573-1612
11. Charbonnier SJ, Gertisser R (2012) Evaluation of geophysical mass flow models using the 2006 block-and-ash flows of Merapi Volcano, Java, Indonesia: Towards a short-term hazard assessment tool. J Volcanol Geotherm Res 231–232:87-108
12. Connor LJ, Connor CB (2006) Inversion is the key to dispersion: understanding eruption dynamics by inverting tephra fallout. In: Mader HM, Coles SG, Connor CB, Connor LJ (eds) Statistics in volcanology, Sp Pub IAVCIE No 1, London: Geological Society.
13. Connor CB, Hill BE, Winfrey B, Franklin NM, LaFemina PC (2001) Estimation of volcanic hazards from tephra fallout. Nat Haz Rev 2:33-42
14. Cortés JA, Palma JL (2013) Petrological INput – Graphical oUtput. https://vhub.org/resources/pingu *webcite*. Accessed 16 Apr 2013
15. Costa A (2013) Hazmap. https://vhub.org/resources/hazmap. Accessed 20 May 2013
16. Courtland LM, Connor C, Connor L, Bonadonna C (2012) Introducing students to numerical modeling of volcanic hazards: the example of Tephra2 on VHub.org. Numeracy 5(2):6 doi:10.5038/1936-4660.5.2.6
17. Crosweller HS, Arora B, Brown SK, Cottrell E, Diligne NI, Ortiz Guerrero N, Hobbs L, Kiyosugi K, Loughlin SC, Lowndes J, Nayembil M, Siebert L, Sparks RSJ, Takarada S, Venzke E (2012) Global database on large magnitude explosive volcanic eruptions (LaMEVE). J Appl Volcanol 1:4 doi:10.1186/2191-5040-1-4
18. Dalbey K, Patra A (2008) Input uncertainty propagation methods and hazard mapping of geophysical mass flows. J Geophys Res 113:1-16

19. Davenport TH, De Long DW, Beers MC (1998) Successful knowledge management projects. Sloan Management Rev 39:43-57
20. Denlinger RP, Iverson RM (2001) Flow of variably fluidized granular masses across three-dimensional terrain: 2. Numerical predictions and experimental tests. J Geophys Res 106:553-566
21. Elmagarmid AK, Samuel A, Ouzzani M (2008) Community-Cyberinfrastructure-Enabled discovery in science and engineering. Comp Sci Eng 10(5):46-53
22. Iverson RM (1997) The physics of debris flows. Rev Geophys 35:245-296
23. Iverson RM, Denlinger RP (2001) Flow of variably fluidized granular masses across three dimensional terrain 1. Coulomb mixture theory. J Geophys Res 106:537-552
24. Johnston EN, Phillips JC, Bonadonna C, Watson IM (2012) Reconstructing the tephra dispersal pattern from the Bronze Age eruption of Santorini using an advection–diffusion model. Bull Volcanol 74:1485-1507 doi:10.1007/s00445-012-0609-x
25. Kanamitsu M, Ebisuzaki W, Woollen J, Yang S-K, Hnilo JJ, Fiorino M, Potter GL (2002) NCEP–DOE AMIP-II Reanalysis (R-2). Bull Am Meteorol Soc 83:1631-1643
26. Klimeck G, McLennan M, Brophy SP, Adams GB III, Lundstrom MS (2008) nanoHUB.org: Advancing Education and Research in Nanotechnology. Comp Sci Eng 10:17-23
27. Levelt PF, van den Oord GHJ, Dobber MR, Mälkki A, Visser H, de Vries J, Stammes P, Lundell JOV, Saari H (2006) The ozone monitoring instrument. IEEE Trans Geosci Remote Sensing 44:1093-1101
28. Levelt PF, Hilsenrath E, Leppelmeier GW, van den Oord GHJ, Bhartia PK, Tamminen J, de Haan JF, Veefkind JP (2006) Science Objectives of the Ozone Monitoring Instrument. IEEE Trans Geosci Remote Sensing 44:1199-1208
29. Longchamp C, Bonadonna C, Bachmann O, Skopelitis A (2011) Characterization of tephra deposits with limited exposure: the example of the two largest explosive eruptions at Nisyros volcano (Greece). Bull Volcanol 73:1337-1352 doi:10.1007/s00445-011-0469-9

30. Lopez T, Carn S, Werner C, Kelly P, Fee D, Kelley P, Doukas M, Pfeffer M, Webley P, Cahill C, Schneider D (2012) Evaluation of Redoubt Volcano's sulfur dioxide emissions by the Ozone Monitoring Instrument. J Volcanol Geotherm Res 259:290-307 doi:10.1016/j.jvolgeores.2012.03.002
31. Lube G, Cronin SJ, Thouret J-C, Surono (2011) Kinematic characteristics of pyroclastic density currents and controls on their avulsion from natural and engineered channels. Geol Soc Am Bull 123:1127-1140
32. Macedonio G, Costa A, Longo A (2005) A computer model for volcanic ash fallout and assessment of subsequent hazard. Comp Geosci 31:837-845
33. Mangeney-Castelnau A, Vilotte J, Bristeau MO, Perthame B, Bouchut F, Simeoni C, Yernini S (2002) Numerical modeling of avalanches based on Saint-Venant equations using a kinetic Scheme. J Geophys Res 108:2537 doi:10.1029/2002JB002024
34. Marzocchi W, Sandri L, Selva J (2010) BET_VH: a probabilistic tool for long-term volcanic hazard assessment. Bull Volcanol 73:705-716 doi:10.1007/s00445-010-0357-8
35. Mastin LG (2011) Plumeria 2.3.1. https://vhub.org/resources/1194. Accessed 16 Apr 2013
36. McLennan M, Kennell R (2010) HUBzero: a platform for dissemination and collaboration in computational science and engineering. Comp Sci Eng 12:48-52
37. Palma JL (2012) Titan2D Viewer. https://vhub.org/resources/titanview. Accessed 20 May 2013
38. Palma JL (2013a) Energy cone. https://vhub.org/resources/econe Accessed 20 May 2013
39. Palma JL (2013b) DEM converter for Titan2D. https://vhub.org/resources/dem4titan *webcite*. Accessed 20 May 2013
40. Patra AK, Bauer AC, Nichita CC, Pitman EB, Sheridan MF, Bursik MI, Rupp B, Webber A, Stinton AJ, Namikawa LM, Renschler CS (2005) Parallel adaptive numerical simulation of dry avalanches over natural terrain. J Volcanol Geotherm Res 139:1-21
41. Pitman EB, Nichita CC, Sheridan M, Bursik MI (2003) Computing granular avalanches and landslides. Phys Fluids 15:3638-3646 Publisher Full Text

42. Rose WI (2011) AshFall: A graduate course in volcanology with substantial meteorological content. https://vhub.org/resources/411 Accessed 16 Apr 2013
43. Rosenberg MJ (2001) E-Learning: Strategies for delivering knowledge in the digital era. New York: McGraw Hill.
44. Savage SB, Hutter K (1989) The motion of a finite mass of granular material down a rough incline. J Fluid Mech 199:177-215
45. Sheridan MF, Macías JL (1995) Estimation of risk probability for gravity-driven pyroclastic flows at Volcan Colima, Mexico. J Volcanol Geotherm Res 66:251-256
46. Shih W-C, Tseng S-S, Yang C-T (2008) Wiki-based rapid prototyping for teaching-material design in e-learning grids. Comput Educat 51:1037-1057
47. Siebert L, Simkin T, Kimberly P (2010) Volcanoes of the World. Berkeley: Smithsonian Institution, University of California Press.
48. Stefanescu ER, Bursik MI, Patra AK (2012) Effect of digital elevation models on Mohr-Coulomb geophysical flow model output. Nat Haz 62:635-656 Publisher Full Text
49. Stinton AJ, Sheridan MF, Patra AK, Dalbey KR, Namikawa LM (2004) Integrating variable bed friction into the TITAN2D mass-flow model: applications to the Little Tahoma Peak Avalanches. Acta Vulcanol 16:153-163
50. Surono JP, Pallister J, Boichu M, Buongiorno MF, Budisantoso A, Costa F, Andreastuti S, Prata F, Schneider D, Clarisse L, Humaida H, Sumarti S, Bignami C, Griswold J, Carn SA, Oppenheimer C, Lavigne F (2012) The 2010 explosive eruption of Java's Merapi volcano – a '100-year' event. J Volcanol Geotherm Res 241–242:121-135
51. Thomas HE, Watson IM, Carn SA, Prata AJ, Realmuto VJ (2011) A comparison of AIRS, MODIS and OMI sulphur dioxide retrievals in volcanic clouds. Geomat Nat Haz Risk 2:217-232
52. Tonini R, Sandri L, Selva J, Marzocchi W (2012) Bayesian event tree for volcanic hazard. https://vhub.org/resources/betvh *webcite*. Accessed 16 Apr 2013
53. Volentik ACM, Connor CB, Connor LJ, Bonadonna C (2009) Aspects of volcanic hazards assessment for the Bataan nuclear power plant, Luzon Peninsula, Philippines. In: Connor CB,

Chapman N, Connor L (eds) Volcanic and Tectonic Hazard Assessment for Nuclear Facilities, New York: Cambridge University Press.

54. Volentik ACM, Bonadonna C, Connor CB, Connor LJ, Rosi M (2010) Modeling tephra dispersal in the absence of wind: Insights from the climatic phase of the 2450 BP Plinian eruption of Pululagua volcano (Ecuador). J Volcanol Geotherm Res 193:117-136

55. Yu B, Dalbey K, Webb A, Bursik M (2009) Numerical issues in computing inundation areas over natural terrains using Savage-Hutter theory. Nat Haz 50:249-267

Chapter 7

The Cost of Living in the Anthropocene

Michael R Gillings[1] and Elizabeth L Hagan-Lawson[2]

[1]Department of Biological Sciences, Macquarie University, Sydney, NSW 2109, Australia

[2]Department of Anthropology, Macquarie University, Sydney, NSW 2109, Australia

ABSTRACT

The most recent epoch, the Holocene, has been a period of relative environmental stability, allowing humans to develop agriculture and establish settlements, culminating in modern civilization. Human activities have now reached such a scale that we are having significant impacts on planetary systems, and these effects are of sufficient magnitude to suggest that we have triggered a new geological epoch, the Anthropocene. Neither climatic nor biogeochemical stability is likely to continue in the Anthropocene, and the Earth systems we rely on to provide a liveable environment for human society are likely to become much less predictable. The stability of our infrastructure, the reliability of our production systems and the liveability of our cities will all be much less certain in the future. More research on the diverse aspects of global change will certainly help to improve predictions on the timing and extent of changes, but will not alter the basic conclusion that global change is upon us. There is now a pressing need for much more interdisciplinary work, addressing such questions as the global societal changes that must accompany responses to environmental change, and dealing with the true economic consequences of a less predictable environment. Conceptualizing the challenges that face humanity under the umbrella of the Anthropocene should allow different disciplines to collaborate and develop strategies for dealing with global change in a coherent and rational manner. Researchers in diverse fields must work together with primary producers, politicians, business interests, policy makers and the public to formulate strategies to minimise or mitigate the risks that face all of humanity over the next centuries. Here we provide a summary of the environmental triggers that are pushing us into the Anthropocene, and outline the consequences of transgressing the boundaries beyond which earth systems are likely to become unstable.

BACKGROUND

The history of Earth is divided into geological periods. The most recent period is the Quaternary, which spans the last 2.5 million years, and corresponds roughly to the time that hominids have occupied the Earth. In turn, periods are subdivided into epochs. For instance, the Quaternary is divided into the Pleistocene and the Holocene epochs. Transitions between Periods or Epochs are defined by major geological or paleontological events that leave a signature in the geological record. Such signatures are caused by significant changes in climate, atmosphere or biota. The best-known example of such a transition is the catastrophic event that ended the Cretaceous period and coincided with the extinction of the dinosaurs.

For the last 10,000 years, humans have been living in the Holocene epoch. This has been an interglacial period of relative environmental stability, where temperature, atmospheric conditions and biogeochemical cycles have exhibited only minor fluctuations

(Dansgaard et al. 1993; Rioual et al.2001; Young and Steffen 2009). This stability allowed humans to develop agriculture and form settled communities, culminating in the complex societies of modern times. The very stability of the Holocene has helped humans to abandon a mobile lifestyle that exploited natural resources, in favour of permanent settlements with complex infrastructure (van der Leeuw 2008).

Modern societies are highly dependent on the physical and electronic links between larger and larger regions, and especially upon the transport of matter and energy. The timeliness and reliability of such systems is dependent on the predictability of environmental conditions. However, our infrastructure and agriculture have been developed to function under the benign conditions that have typified the Holocene, and the last 10,000 years. It is thus of considerable concern that human activities have begun to alter the Earth's climate and biogeochemistry, threatening to plunge us into an era of unpredictability in all Earth systems. It is not certain that we will be able to adapt quickly enough to accommodate these changes. The scale and speed of change has led to the proposal that we are moving into a new epoch, termed the Anthropocene, or literally, the 'Age of Man' (Crutzen 2002; Steffen 2006; Steffen et al. 2007).

So what is the evidence that we should recognize the current era as a new geological epoch? What are the key atmospheric, biotic and environmental changes that have already occurred, and what changes can we expect in the future? Can we actually predict the unpredictable? The destabilization of key planetary systems may have catastrophic consequences for humanity, and yet current policies of social and economic development often operate in an arena of selective blindness to looming planetary disasters (Stern 2007). Strategies for policy and governance must be far more responsive to the current and ongoing destabilization of planetary scale processes if humanity is to deal successfully with the predicted rate of environmental change (Biermann 2012). Furthermore, the science of global change must be communicated in language that is concise and clear, such that non-specialists can understand it, yet still be accurate. Here we review recent literature on the Anthropocene, and the interactions between the various phenomena contributing to global change.

The History of the Anthropocene

The idea that human activities have the power to affect Earth's systems was recognized in the late 1800s by the geologist Antonio Stoppani, who coined the term 'Anthropozòic Era' to highlight this conclusion (Crutzen 2002). The term 'Anthropocene' was first used by Eugene Stoermer, and subsequently popularized by the atmospheric chemist and Nobel Prize laureate Paul Crutzen (Crutzen2002). The Anthropocene may soon become an official epoch, since a proposal to formalise it was made to the Stratigraphy Commission of the Geological Society of London in 2008, resulting in serious consideration for its acceptance as an epoch in the geological time scale (Zalasiewicz et al.2008; Zalasiewicz et al. 2010).

Crutzen dates the beginning of the Anthropocene to the late 18th century when atmospheric concentrations of carbon dioxide and methane began to rise significantly. Other writers have suggested that the Anthropocene could be dated to as long as 8,000 years ago, driven by clearing of forests and agriculture (Ruddiman 2003; Ruddiman 2013). However, the 18th century date seems a reasonable compromise, since this date coincides with the invention of the steam engine and the industrial revolution, and corresponds to increased carbon emissions by humans.

Biotic and Geochemical Markers of the Anthropocene

The most compelling argument for the erection of the Anthropocene as a new epoch relies on projecting ourselves forward in time, and asking if future geologists will be able to detect a stratigraphic boundary that defines a transition from the Holocene (Zalasiewicz et al. 2011; Gale and Hoare 2012; Brown et al. 2013). To this end, a number of key processes have been identified as 'Planetary Boundaries', representing thresholds below which humanity can safely operate, and beyond which the stability of planetary-scale systems cannot be relied upon (Rockstrom et al. 2009a; Rockstrom et al. 2009b). Transgressing these boundaries may leave a clear record in the geological column.

Nine distinct planetary boundaries have been suggested, each based on the unacceptable economic, social and environmental risks that humanity faces should they be crossed. For each of these boundaries attempts have been made to estimate the tipping points, beyond which abrupt and irreversible environmental changes might occur. There is uncertainty about many of these boundaries, because we do not have enough data and we lack a comprehensive understanding of the complex feedback mechanisms that maintain resilience and stability in biophysical systems (Rockstrom et al. 2009a). Nevertheless, we should operate from the precautionary principle that abrupt and catastrophic changes in earth systems are likely if humanity continues its current scale of activities.

The key planetary boundaries, and their potential consequences, are as follows:

Climate Change

Humans have had a significant effect on global carbon cycling. Emissions of carbon dioxide into the atmosphere have raised its concentration to a level higher than at any time in the last 800,000 years. Carbon isotope chemistry was used to mark the boundary between the Paleocene and Eocene epochs, so a carbon signature also seems appropriate to mark a transition to the Anthropocene (Steffen et al.2011a).

Thirty years ago there were suggestions that the Earth was warming, based on the known linkage between global mean temperature and the atmospheric concentration of carbon dioxide. Despite political opposition and the efforts of climate sceptics, this conclusion is now certain (Agnihotri and Dutta 2013; IPCC 2007). The boundary for deleterious climate change forced by accumulation of carbon dioxide was set at 350 ppm, and this boundary has already been transgressed (Steffen et al.2007; Rockstrom et al. 2009a).

Climate change has the potential to affect every aspect of our economy and society. Extreme weather events, such as hurricanes, heatwaves and droughts will increase in intensity and frequency over the next decades (Repetto and Easton 2010; Coumou and Rahmstorf 2012). Around the globe, weather related events already cause more casualties than other catastrophes (Bouwer 2012), and have serious human health consequences (Knowlton et al. 2011). In addition to the direct human toll, extreme weather disrupts or destroys infrastructure. For instance, in October 2012, Hurricane Sandy resulted in the cancellation of almost 20,000 airline flights and the closure of bus, rail and ferry services (Kaufman et al. 2012). Total damages from Sandy were estimated to lie between 30 and 50 billion dollars (Tollefson 2012). Globally, economic losses caused by extreme weather events are rapidly increasing, and will continue to rise over the next decades (Bouwer 2012). Strategic decision making by businesses should anticipate and incorporate the need for organizational adaptation in the face of this rapidly changing climate (Linnenluecke et al. 2012).

Predicting the impacts of climate change is difficult because the likely effects are not uniform in either location or direction. For instance, weather in north-western Europe is dependent on the effect of the warm North Atlantic Drift. Changes in this thermohaline circulation caused by alterations in rainfall patterns could, ironically, lower temperatures by as much as 10°C across NW Europe, and such changes might be irreversible (Rahmstorf 2000).

The last decade has seen an increase in the frequency and severity of heatwaves and droughts (Coumou and Rahmstorf 2012), both of which have serious implications for food production, human health, and infrastructure (Fischer and Schar 2010; Anderson and Bell 2011; Battisti and Naylor2009). There will also be a substantial increase in the frequency of intense tropical cyclones and the attendant damage to cities and coastal settlements. Category 4 and 5 storms are expected to double in frequency by the end of the century (Bender et al. 2010; Knutson et al. 2010). There is also good evidence that the frequency of great floods has increased significantly over the last 100 years (Milly et al. 2002). Such floods damage settlements, impact agriculture and promote the spread of waterborne diseases (Few 2012).

Rising global temperatures lead to proportional rises in sea level, due to thermal expansion of the oceans and melting of ice fields (Hodgson 2011; Nicholls and Cazenave 2010). Rates of sea level rise are currently 3.4 mm/year with an estimate of a total rise of between 0.5 and 1.4 m by 2100 (Rahmstorf 2007; Rahmstorf 2010; Schaeffer et al. 2012). Impacts on humans will include inundation of low lying areas and loss of islands with low elevation. Although such effects are restricted to coastal areas, these are often highly productive ecosystems with a consequently high population density. Densely populated cities such as Tokyo, Shanghai and Bangkok exhibit the additional problem of land subsidence, exacerbating the effects of rising sea levels (Nicholls and Cazenave 2010). Estimates suggest that some 10 to 50 million people in the Southeast Asian region alone will become refugees from sea level rise (Wetzel et al. 2012). Many of the regions most at risk also have an inherently low capacity for adaptation to rising sea levels (Nicholls and Cazenave 2010), and the greatest impacts will disproportionately affect those least responsible for climate change.

Worldwide, populations in the hazard zone for 1000 year storm surges are estimated to reach 400 to 600 million by 2050. As a contemporary example, in October 2012 Hurricane Sandy caused record storm surges in New Jersey, NY City, Connecticut, and Rhode Is. Other significant effects of sea level rise include loss of wetlands, erosion and salt-water intrusion into freshwater systems (Nicholls and Tol 2006), thus having flow-on effects on other planetary boundaries such as biodiversity, land use and fresh water resources.

Further effects of climate change include an increase in the frequency and severity of fires, loss of alpine habitats, movement of vegetation zones and disruption to agriculture (Hughes 2003). Declines in crop yields due to climate change are already apparent, with a global drop in wheat and maize production of 5.5 and 3.8% respectively, despite improvements in agricultural technology over the last 30 years (Lobell et al. 2011). Modelling suggests that climate change will result in yield declines of 30-80% in US production of corn and soybeans by the turn of the century (Schlenker and Roberts2009). Long term global trends are more difficult to predict because effects differ significantly from region to region (Piao et al. 2010; Schlenker and Lobell 2010). World production of food may remain stable, but declines in yield will disproportionately affect developing nations, resulting in higher prices, increased risks of food shortages and increasing inequity in food distribution (Parry et al.2004; Lobell et al. 2008).

Ocean Acidification

Nearly one third of the carbon dioxide released by anthropogenic activity is absorbed by the oceans. But for this fact, current atmospheric CO_2 concentrations would be higher than they already are. However, CO_2 uptake lowers the pH and alters the chemical balance of the oceans, in particular the solubility of calcium salts (Doney et al. 2009). This phenomenon is called ocean acidification, and is occurring at a rate faster than at any time in the last 300 million years (Hönisch et al. 2012). We are now at risk of transgressing the pH boundary where the calcium carbonate and aragonite skeletons of marine life will simply dissolve (Rockstrom et al. 2009a). It is a sobering thought that the great Permian extinction event, where 96% of species were lost, was associated with ocean acidification (Barnosky et al. 2011; Harnik et al. 2012).

The economic costs of ocean acidification encompass a number of biological effects, including decalcification of the skeletons of molluscs, echinoderms, foraminifera, coralline red algae and tropical corals, and effects on reproduction of plankton, molluscs and echinoderms (Doney et al. 2009; Lurling and De Senerpont Domis 2013; Schlegel et al. 2012). Many of these groups contain keystone species, whose loss will permanently alter marine ecosystems.

Corals reefs harbor a significant proportion of marine diversity. Ocean acidification threatens to dissolve coral skeletons and convert reefs into algal dominated systems (Hoegh-Guldberg et al. 2007; Pandolfi et al. 2011). The economic impact on ecosystem services provided by reefs are difficult to assess (Stoeckl et al. 2011), but there will be serious consequences for fisheries, tourism and coastal protection (Hoegh-Guldberg et al. 2007). Estimates suggest that current global loss in shellfish production due to ocean acidification amounts to some 6 billion dollars (US) per annum, and that this could rise to $100 billion, based on increased demand in the Asia-Pacific region (Narita et al. 2012). Degradation of marine habitats, loss of ecosystem services and the resultant decline in harvests are likely to have disproportionate effects in some regions, particularly those with high dependency on marine resources (Cooley and Doney 2009). The number of peer-reviewed analyses on the economic impact of ocean acidification is still small, but is likely to increase sharply, given the rapid rise in research papers now being published in this area. Appropriate methods for estimating costs have been identified and applied to ecosystem services in Norwegian waters. There were both positive and negative effects of acidification, but overall there was a large net negative outcome (Armstrong et al.2012). In the final analysis, protection of the ocean may be more important than protection of atmosphere or land because it stores more carbon, mediates climate variability and provides essential ecosystem services (Steffen et al. 2011b).

Ozone Depletion

The original appearance of the ozone layer allowed the emergence of life on land, because ozone in the stratosphere filters out ultraviolet radiation from the sun. Human use of chlorofluorocarbons and other compounds has depleted ozone, and resulted in the thinning of the ozone layer in the polar stratosphere. This exposes humans to health risks and has negative effects on marine ecosystems. Ozone depletion is also the major cause of changes in the pattern of atmospheric circulation in the southern hemisphere (Portmann et al. 2012). Climate models suggest that ozone depletion led to a significant increase in rainfall across SE South America during the last 40 years of the 20th century. Any recovery of the ozone layer in the future may thus lead to a drying of this region, with consequent adverse effects on agriculture and economies (Gonzalez et al. 2013).

Recognition of the ozone depletion problem led to an international agreement (the Montreal Protocol), and as a consequence, ozone-depleting halocarbons have declined from their peak in 1992–1994. Globally, we are now unlikely to pass the tipping point for ozone depletion, largely based on acceptance of key scientific findings, and the resulting international agreements (Velders et al. 2007; Mäder et al. 2010). Although ozone depleting chemicals will take many decades to degrade, informed science and political will have changed human behaviour so successfully as to avoid transgressing the tipping point for ozone depletion (Rockstrom et al. 2009a). However, the reduction of halocarbon emissions will not solve the problem entirely, since anthropogenic nitrous oxide emission also affects ozone levels (Portmann et al. 2012).

Atmospheric Aerosol Loading

Aerosols are particles suspended in the atmosphere. Since the industrial revolution the concentration of aerosols has doubled (Tsigaridis et al. 2006). There is considerable evidence that aerosols alter the way precipitation forms from clouds, and hence human emissions of aerosols have the potential to significantly change rainfall and weather patterns (Rockstrom et al. 2009a). Simulations suggest that aerosol loading of the atmosphere will change the timing, intensity and extent of the

South Asian monsoon, affecting rainfall over East Asia (Lau et al. 2006; Lewis et al. 2011). Thus one of the most densely populated regions of the planet will suffer most from unpredictable rainfall. Aerosols may also alter the distribution of primary production in marine ecosystems, either enhancing or lowering production dependent on the metal content of the aerosol (Paytan et al. 2009).

Some 80% of populations in Asia are exposed to aerosol concentrations that exceed WHO guidelines (Carmichael et al. 2009). Exposure to the particulate matter in polluted air causes cardiac and respiratory disease, amounting to some 500,000 deaths per year (Nel 2005). The combined health costs due to particulate pollution in Beijing amount to between US$1.6 and 3.6 billion annually, equivalent to over 6% of this cities gross domestic product (Zhang et al. 2007). Globally, the health impacts and consequent economic costs of aerosol emissions are estimated to lie between US$120 and 510 billion annually (Selin et al. 2011). Aerosols also affect other organisms, including crops, forests and aquatic animals, primarily through the acid rain generated by nitrogen and sulphur oxides. Because aerosols are composed of diverse compounds with different origins, half-lives and potential impacts, it is difficult to predict their effects with any certainty, although some progress is being made (Jimenez et al. 2009).

Phosphorus and Nitrogen Cycles

Phosphorus and nitrogen are key elements for growth of living things. Humans have successfully manipulated their availability to improve the productivity of agricultural systems. Prior to the industrial era, the vast majority of biologically available nitrogen was fixed from atmospheric nitrogen via bacterial activity. By tapping into fossil fuels and as a result of agricultural activities, humans now fix more atmospheric nitrogen into biologically available forms than the quantity fixed by all natural processes combined (Rockstrom et al. 2009a; Gruber and Galloway 2008). Industrial fixation of nitrogen is accomplished via the energy intensive Haber-Bosch process, while other human-driven processes such as cultivation of legumes, fossil fuels and burning of biomass add to the total. The planetary boundary for the nitrogen cycle has already been passed (Rockstrom et al. 2009a).

Nitrogen flux through the biosphere is primarily a biological process, while phosphorus availability arises slowly through geological weathering. Humans sidestep the phosphorus bottleneck by mining and distribution of fertilizer onto agricultural lands, thus inadvertently increasing the flow of phosphorus into the oceans. While the planetary boundary for phosphorus is unlikely to be passed in the near future (Rockstrom et al. 2009a), the production of phosphorus for fertilizer will peak in 2030, before peak demand occurs. In a world where phosphorus supply is limited, there are serious implications for global food security (Cordell et al. 2009).

Increasing levels of phosphorus and nitrogen in soils and waters causes soil acidification, eutrophication of waterways and is associated with harmful algal blooms (Anderson et al. 2008; Bennett et al. 2001). Eutrophication affects waterfront real estate values, recreational water use, freshwater availability and expenditure on species recovery programs. In the US, the combined costs arising from eutrophication have been estimated as US$2.2 billion per year (Dodds et al. 2008). In England and Wales, the damage costs of eutrophication are in the range of US$105-160 million per year (Fankhauser 2009). Excess phosphorus flowing into the oceans also causes oceanic anoxia, associated with mass extinction events (Barnosky et al. 2011; Handoh and Lenton 2003). Guidelines and protocols for lowering the release of nutrients into ecosystems must be widely adopted to lessen the economic impact of eutrophication (Lewis et al. 2011; Fulweiler et al. 2012).

Global Freshwater Use

Humans die in several days if water is not available. Because of this, access to fresh water is regarded as a universal human right (Füssel et al. 2012). The water bodies that we depend on also connect the atmosphere, soils, and the oceans, and are indispensable for the functioning of terrestrial ecosystems. However, freshwater systems can no longer be thought of as being naturally regulated. Human use of freshwater and our control of river flows have fundamentally altered the water cycle at both local and regional scales (Meybeck 2003). Extraction of water by humans results in 25% of rivers running dry before they reach the ocean (Falkenmark and Molden 2008; Molden 2007). Further, water bodies are subject to a variety of environmental insults including salinization, acidification, and contamination with chemical and microbial pollutants (Meybeck 2003).

Fresh water resources are running out. Many populations live under water stress, and 80% of the world's population face threats to water security (Vorosmarty et al. 2000; Vorosmarty et al. 2010). Limitations on water availability have led to the concept of 'peak water', split into three components: renewable freshwater flow; non-renewable groundwater resources; and peak 'ecological' water. Once the damage cost of any additional human use of water exceeds the benefits, we have passed the point of peak ecological water. Peak water use for at least one of these three components has been passed for major river basins and aquifers all across the globe (Gleick and Palaniappan 2010).

The uncertainty of climate change places the predictability and reliability of water resources in doubt (Piao et al. 2010; Füssel et al. 2012), and the economic impacts scale with climate change. Damage costs for North America are estimated at about US$3 billion for a 1°C increase in global mean temperature, scaling to US$7 - 16 billion at 2.5°C. Globally, costs to water resources are estimated at some US$47 - 84 billion under a 4°C temperature increase (Tol 2002).

In addition to the central role of water in food production, aquatic systems harbour valuable biodiversity and provide essential ecosystem services (Vorosmarty et al. 2010; Wilson and Carpenter1999). Increasing pressure on water availability will have serious economic consequences for food production and human health. Because of the transnational nature of rivers and water bodies, action to protect this resource requires inter-regional and international agreements (Falkenmark and Lundqvist 1998). Adaptive responses to less reliable and diminishing water resources will require critical improvements in governance (Hill 2013).

Land System Change

As human population increases, there is more pressure to convert landscapes into croplands, pastures and urban landscapes. Expansion of agriculture increases the risk of undermining sustainability by threatening biodiversity and affecting both climatic and hydrological cycles (Pielke et al. 2003; Sterling et al. 2013). Changes to land cover driven by deforestation or urbanization interact with climate and temperature, in some cases forcing climate change to the same degree as carbon dioxide emissions. Land use change complicates global climate predictions, but its inclusion may improve modelling at more regional levels (Avila et al. 2012).

Some 12% of the land surface is now under cultivation (Foley et al. 2005), and the suggested planetary boundary of 15% is likely to be surpassed in the near future (Rockstrom et al. 2009a). The current reserves of land suitable for conversion cropping are predicted to be exhausted some time between 2020 and 2050 (Lambin and Meyfroidt 2011). Consequently, some expansion is likely to be forced onto more marginal land, creating further unpredictable feedbacks on biodiversity, hydrology and climate. Additional land use change will be forced upon us by flooding of low lying land during sea level rise. Conversion of dryland by such flooding, and the subsequent cost to ecosystems amount to US$14 billion and $41 billion per year respectively (Fankhauser 1995).

Conversion of biomes to undesirable states, for example forest to desert, can occur rapidly and unpredictably when thresholds are crossed. For instance, small increases in the clearing of Amazonian rainforest may cause an irreversible conversion into semi-arid savannah (Rockstrom et al. 2009a; Foley et al. 2005). Attention to minimising land degradation, conservation of irrigation water and protection of the most productive croplands from urban development will partially mitigate the problem. Similarly, the development of agricultural practices that mimic complex natural ecosystems will minimise the impact of land use (Ericksen et al. 2009). Improved efficiency of land use has dual outcomes of increased agricultural production and forest conservation (Lambin and Meyfroidt 2011). However, continuing human effects on land systems are likely to combine with climate change to convert land systems into less desirable states.

Loss of Biodiversity

Transitions between major geological eras are often characterized by loss of species. When more than 75% of species are lost in a geologically short time period, this is known as a mass extinction. There have been five such mass extinction events over the last 540 million years. In modern times, global biodiversity is declining, via species extinction, at rates that are orders of magnitude higher than might be expected from the fossil record. The rate of decline is not slowing, despite recognition of the problem (Butchart et al. 2010). This has led some scientists to suggest that we are currently in a sixth mass extinction, fully comparable with those in the geological past. Recent analysis suggests that the current loss of species, while dramatic, does not yet qualify as a mass extinction. However, ongoing losses, particularly of species currently recognized as critically endangered, would push us into a state of mass extinction similar in magnitude to the 'big five'. Recovery from such catastrophic species loss would probably take millions of years (Barnosky et al. 2011). It is a sobering thought that the causes of past mass extinctions are eerily similar to key environmental pressures now facing humanity. These include ocean acidification, climate change, changes in atmospheric CO_2 levels, loss of habitat, and deep water anoxia (Barnosky et al. 2011; Harnik et al. 2012).

Extinction of species and loss of biodiversity have multiple impacts on humanity. There is now consensus that biodiversity increases the efficiency of communities in key ecosystem processes, improves the stability of ecosystem functions, and that diverse communities are more productive. Loss of biodiversity has direct economic effects including decreasing the yield in commercial crops, fodder, forestry and fisheries (Cardinale et al. 2012). Over-exploitation of natural resources puts further pressure on biodiversity. For instance, there has been a significant decline in fisheries catches since 1990, despite a doubling of effort (Worm and Branch 2012).

There are numerous conceptual issues surrounding the costs of saving species versus the benefits. Extinction is irreversible, and the effects of individual species loss are uncertain, so the economic costs cannot be accurately estimated. It is difficult to place a value on an ecosystem or a species, but rough estimates based on usage and existence can be used to calculate that a global increase in temperature of 1°C will cost US$50 billion per year in terms of lost biodiversity and resultant ecosystem services (Tol 2002).

The existence value of a species is essentially unmeasurable (Bishop 1978). Further, the effects of losing even one species from an ecosystem are difficult to predict. The magnitude of effects on ecosystem functions increases in a non-linear manner as ever more biodiversity is lost, in part because the majority of species depend on symbioses with other species. Simplification of ecosystems by loss of species may alter their functioning, such that the net effect might be equivalent to effects wrought by ocean acidification and climate change (Reich et al. 2012). Consequently, loss of biodiversity drives changes in ecosystem productivity, decreasing stability and resilience (Hooper et al. 2012).

Chemical Pollution

Humans synthesize and distribute an enormous variety of xenobiotic compounds, with some 80,000 chemicals available in the marketplace. We also mine and disseminate various toxic compounds. In the past, this has resulted in local or regional pollution with heavy metals, synthetic organic compounds and radioactive materials. The extent of such activity means that signatures of chemical pollution are now evident at a global scale, and will be detectable in the geological record.

Because of the diversity of potential pollutants and uncertainty as to their effects, it is difficult to set a universal planetary boundary for chemical pollutants (Rockstrom et al. 2009a). Since chemical pollution encompasses diverse compounds in diverse habitats, there are no universal estimates of the costs of anthropogenic pollution, but rather, key studies dealing with single types of pollutants in particular regions. The global costs of air pollution by nitrogen and sulfur oxides were estimated at US$15 billion per year (Fankhauser 1995), excluding the costs of acid deposition. Clearly this estimate overlaps with the costs of aerosols, of which these oxides are major components. Useful global estimates could be made from the clean-up costs for legacy sites and the costs of pollutant capture, but the ongoing and diffuse nature of chemical pollution makes economic calculations problematic.

We should be particularly concerned about pollutants that have direct biological effects at low concentrations, such as endocrine disrupting chemicals (Diamanti-Kandarakis et al. 2009) and antibiotics (Gillings 2013; Gillings and Stokes 2012), since these have the potential to alter population assemblages and affect biodiversity. There have been successful bans on particular organic pollutants, such as dioxins, PCBs, DDT, and other pesticides. These bans have been based on investigations into their effects on humans and other organisms. Assembling equivalent information for the enormous diversity of compounds currently generated by humans is a daunting task, and is complicated by the potential for interactions between compounds. Certainly pollutants have effects on both ecosystem functioning and on human welfare. Exposure of children to low concentrations of compounds with neurotoxic effects may be leading to a global epidemic of developmental disorders (Grandjean and Landrigan 2006; Grandjean et al. 2008), and many pollutants have the potential for bioaccumulation at higher trophic levels (Arnot and Gobas 2004).

CONCLUSIONS

Humans are having significant effects at a global scale, and the magnitude of these effects is sufficient to suggest we are entering a new geological epoch, the Anthropocene. This new epoch will be characterized by a shift away from the stable conditions that have typified the last 10,000 years, and usher in a period of unpredictability that threatens to destabilise human economics and society. It is clear that the comparative stability that has typified the last 10,000 years will not continue into this century, and that we face a future where business as usual will not be possible.

Attempting to estimate the absolute costs of living in the Anthropocene is difficult, largely because the potential magnitude of environmental effects are hard to quantify, as are the flow-on costs of unpredictability in weather, nutrient cycling and ecosystem services. Here we have dealt with these potential costs via the framework of the planetary boundaries defined by Rockstrom et al. (Rockstrom et al. 2009b), however this framework does not encompass all the costs inherent in the transition to the Anthropocene. For instance, the contributions of cultural heritage such as landscape aesthetics, the cultural significance of particular sites, or the intrinsic value of outdoor recreation are not well integrated into analyses of ecosystem services (Daniel et al. 2012). A number of groups are actively working to incorporate such values into ecosystem assessments (Tengberg et al. 2012; Chan et al.2012).

The terms 'ecology' and 'economics' both use the same Greek root as a prefix (oikos?=?home), and it is surprising that there has not been more cross-disciplinary work between the two fields. Indeed, the value of ecosystem services, essential for human welfare, was not estimated until 1997 (Costanza and Déarge 1997). The value of such services was placed in the range of US$16-54 trillion dollars, most of which lay outside the market. By comparison, at that time the global gross national product was US$18 trillion.

Since 1997, there has been more attention paid to valuing the world's natural capital, and examining the trade-offs between human welfare and the environment (Farber et al. 2002; Fisher et al. 2011). Particular attention has been paid to biodiversity and conservation (Naidoo et al. 2008), and the costs associated with loss of forest ecosystem services (Chiabai et al. 2011). There has also been critical analysis of the interactions between ecosystem services, policy and economics, and linking these to the costs of human activities (Balmford et al. 2011; Farley 2012; Kinzig et al. 2011). It is clear that much more attention needs to be paid to the real costs resulting from human exploitation of natural resources, and this in turn will require closer collaboration between environmental scientists and economists. Finally, while science can inform and predict, its practitioners will increasingly have to deal with the more difficult issues of value, amenity and ethics (Seidl et al. 2013).

Dealing with the instability and environmental consequences of the Anthropocene will require cooperation and coordination between scientists, policy makers, governments and the general public. However, governance is required at time scales beyond past experience, because the earth systems we are affecting operate at a time scale that is mismatched with human decision making and our economic systems (Steffen et al. 2007). The time lag between cause and effect for complex earth systems mean that likely effects are already mortgaged into the future, and preventing the causes now will not stop the changes that are already in train (Steffen et al. 2011a). Nevertheless, whole society and global responses are needed to prevent even more catastrophic changes to earth systems. Such responses must include population control, greatly reduced consumption of resources, sustainable energy generation, conservation of the natural world and generally better management of all human activities. These actions are essential if we are to maintain human welfare on a less predictable planet (Tickell 2011).

Effective management of earth systems will require changes to both national and international organizations (Biermann et al. 2012), and to aid this process an Earth System Governance Project has been proposed (Biermann et al. 2010). Clearly, adaptive responses have to be examined at local, regional and global scales, and further, because the concept of planetary boundaries crosses environmental, social and economic systems, communication between different fields must be improved, to adequately represent each discipline's perspectives (Nilsson and Persson 2012; Veldkamp et al. 2011). The Anthropocene concept offers the ripe opportunity, and indeed, the necessity, for truly transdisciplinary research that transcends sociology, politics, economics and environmental science (Seidl et al. 2013; Hoban and Vernesi 2012; Palsson et al. 2013). The planetary boundaries central to the Anthropocene concept imply social tipping points that accompany the planetary tipping points, creating a research opportunity for fusing the work of those interested in politics, governance, human psychology and ecology (Biermann 2012). The true economic costs of the Anthropocene are yet to be properly quantified for any one of the planetary boundaries, and this is also an area that needs immediate attention from diverse research standpoints. Finally, solutions based on geoengineering rather than attitudinal and social change need to be examined in terms of their ecological, social and economic cost (Galaz 2012).

Many human civilizations have collapsed in the past, usually due to complex combinations of environmental, social and economic factors (Butzer and Endfield 2012; Diamond 1994). However, modern society has the advantage of hindsight, and knows the kinds of disasters that have overtaken previously successful civilizations. We owe it to future generations to not repeat the mistakes made in the past (Lawson 2011), and to invest in the future of our planet now, rather than pay a higher price later. Acceptance of the risks we face is a significant hurdle to overcome, since individuals and institutions are in a state of collective cognitive dissonance (Steffen et al. 2011a), where unpalatable facts are actively ignored, or worse, distorted and ridiculed. The folly of using growth as a measure of success needs to be more widely appreciated, since continuous growth in human activity is inimical to the natural processes upon which human welfare depends (Kosoy et al. 2012). Current policies tend to ignore intangible and long-term costs, including those of environmental degradation. Unless we consider the true costs of consumption by placing real values on natural capital, we run the unacceptable risk of squandering our inheritance.

AUTHORS' CONTRIBUTIONS

MRG and ELH conceived and designed the manuscript, assembled and reviewed the literature. MRG wrote the manuscript drafts, and ELH provided critical revisions. Both authors have read and approved the final manuscript.

ACKNOWLEDGEMENTS

Responsible editor: Michael Schaepman

REFERENCES

1. Agnihotri R, Dutta K (2013) Anthropogenic Climate Change: Observed Facts, Projected Vulnerabilities and Knowledge Gaps. In: Sinha R, Ravindra R (eds) Earth System Processes and Disaster Management, Berlin Heidelberg: Springer. pp 123-137
2. Anderson GB, Bell ML (2011) Heat waves in the United States: mortality risk during heat waves and effect modification by heat wave characteristics in 43 US communities. Environ Health Perspect 119(2):210-218
3. Anderson DM, Burkholder JM, Cochlan WP, Glibert PM, Gobler CJ, Heil CA, Kudela RM, Parsons ML, Rensel JEJ, Townsend DW, et al. (2008) Harmful algal blooms and eutrophication: Examining linkages from selected coastal regions of the United States. Harmful Algae 8(1):39-53
4. Armstrong CW, Holen S, Navrud S, Seifert A (2012) The Economics of Ocean Acidification–a scoping study. Norway: Frams Center.
5. Arnot JA, Gobas FAPC (2004) A food web bioaccumulation model for organic chemicals in aquatic ecosystems. Environ Toxicol Chem 23(10):2343-2355
6. Avila FB, Pitman AJ, Donat MG, Alexander LV, Abramowitz G (2012) Climate model simulated changes in temperature extremes due to land cover change. J Geophys Res 117(D4):D04108.
7. Balmford A, Fisher B, Green R, Naidoo R, Strassburg B, Kerry Turner R, Rodrigues A (2011) Bringing Ecosystem Services into

the Real World: An Operational Framework for Assessing the Economic Consequences of Losing Wild Nature. Environ Resour Econ 48(2):161-175

8. Barnosky AD, Matzke N, Tomiya S, Wogan GOU, Swartz B, Quental TB, Marshall C, McGuire JL, Lindsey EL, Maguire KC, et al. (2011) Has the Earth/'s sixth mass extinction already arrived? Nature 471(7336):51-57
9. Battisti DS, Naylor RL (2009) Historical Warnings of Future Food Insecurity with Unprecedented Seasonal Heat. Science 323(5911):240-244
10. Bender MA, Knutson TR, Tuleya RE, Sirutis JJ, Vecchi GA, Garner ST, Held IM (2010) Modeled impact of anthropogenic warming on the frequency of intense Atlantic hurricanes. Science 327:454-458
11. Bennett EM, Carpenter SR, Caraco NF (2001) Human Impact on Erodable Phosphorus and Eutrophication: A Global Perspective. BioScience 51(3):227-234
12. Biermann F (2012) Planetary boundaries and earth system governance: Exploring the links. Ecol Econ 81:4-9
13. Biermann F, Betsill MM, Vieira SC, Gupta J, Kanie N, Lebel L, Liverman D, Schroeder H, Siebenhüner B, Yanda PZ, et al. (2010) Navigating the anthropocene: the Earth System Governance Project strategy paper. Curr Opin Environ Sustain 2(3):202-208
14. Biermann F, Abbott K, Andresen S, Bäckstrand K, Bernstein S, Betsill M, Bulkeley H, Cashore B, Clapp J, Folke C (2012) Navigating the Anthropocene: Improving Earth System Governance. Science 335(6074):1306-1307
15. Bishop RC (1978) Endangered Species and Uncertainty: The Economics of a Safe Minimum Standard. Am J Agric Econ 60(1):10-18
16. Bouwer LM (2012) Projections of Future Extreme Weather Losses Under Changes in Climate and Exposure. Risk Anal.
17. on-line early doi:10.1111/j.1539-6924.2012.01880.x.
18. Brown AG, Tooth S, Chiverrell RC, Rose J, Thomas DSG, Wainwright J, Bullard JE, Thorndycraft VR, Aalto R, Downs P (2013) The Anthropocene: is there a geomorphological case? Earth Surf Processes Landforms 38(4):431-434

19. Butchart SHM, Walpole M, Collen B, van Strien A, Scharlemann JPW, Almond REA, Baillie JEM, Bomhard B, Brown C, Bruno J, et al. (2010) Global Biodiversity: Indicators of Recent Declines. Science 328(5982):1164-1168
20. Butzer KW, Endfield GH (2012) Critical perspectives on historical collapse. Proc Natl Acad Sci 109(10):3628-3631
21. Cardinale BJ, Duffy JE, Gonzalez A, Hooper DU, Perrings C, Venail P, Narwani A, Mace GM, Tilman D, Wardle DA, et al. (2012) Biodiversity loss and its impact on humanity. Nature 486(7401):59-67
22. Carmichael GR, Adhikary B, Kulkarni S, D'Allura A, Tang Y, Streets D, Zhang Q, Bond TC, Ramanathan V, Jamroensan A, et al. (2009) Asian Aerosols: Current and Year 2030 Distributions and Implications to Human Health and Regional Climate Change. Environ Sci Technol 43(15):5811-5817
23. Chan K, Satterfield T, Goldstein J (2012) Rethinking ecosystem services to better address and navigate cultural values. Ecol Econ 74:8-18
24. Chiabai A, Travisi C, Markandya A, Ding H, Nunes P (2011) Economic Assessment of Forest Ecosystem Services Losses: Cost of Policy Inaction. Environ Resour Econ 50(3):405-445
25. Cooley SR, Doney SC (2009) Anticipating ocean acidification's economic consequences for commercial fisheries. Environ Res Lett 4(2):024007
26. Cordell D, Drangert J-O, White S (2009) The story of phosphorus: Global food security and food for thought. Glob Environ Chang 19(2):292-305
27. Costanza R, Déarge R, Groot RS, Farber S, Grasso M, Hannon B, Limburg K, Naeem S, O'Neill R (1997) The Value of the World's Ecosystem Services and Natural Capital. Nature 387:253-260
28. Coumou D, Rahmstorf S (2012) A decade of weather extremes. Nature Climate Change 2(7):491-496.
29. Crutzen PJ (2002) Geology of mankind. Nature 415:23
30. Daniel TC, Muhar A, Arnberger A, Aznar O, Boyd JW, Chan KMA, Costanza R, Elmqvist T, Flint CG, Gobster PH, et al. (2012) Contributions of cultural services to the ecosystem services agenda. Proc Natl Acad Sci 109(23):8812-8819

31. Dansgaard W, Johnsen SJ, Clausen HB, Dahl-Jensen D, Gundestrup NS, Hammer CU, Hvidberg CS, Steffensen JP, Sveinbjornsdottir AE, Jouzel J, et al. (1993) Evidence for general instability of past climate from a 250-kyr ice-core record. Nature 364(6434):218-220
32. Diamanti-Kandarakis E, Bourguignon J-P, Giudice LC, Hauser R, Prins GS, Soto AM, Zoeller RT, Gore AC (2009) Endocrine-Disrupting Chemicals: An Endocrine Society Scientific Statement. Endocr Rev 30(4):293-342
33. Diamond J (1994) Ecological collapses of past civilizations. Proc Am Philos Soc 138(3):363-370
34. Dodds WK, Bouska WW, Eitzmann JL, Pilger TJ, Pitts KL, Riley AJ, Schloesser JT, Thornbrugh DJ (2008) Eutrophication of US freshwaters: analysis of potential economic damages. Environ Sci Technol 43(1):12-19
35. Doney SC, Fabry VJ, Feely RA, Kleypas JA (2009) Ocean Acidification: The Other CO2 Problem. Ann Rev Mar Sci 1(1):169-192
36. Ericksen PJ, Ingram JSI, Liverman DM (2009) Food security and global environmental change: emerging challenges. Environ Sci Policy 12(4):373-377
37. Falkenmark M, Lundqvist J (1998) Towards water security: political determination and human adaptation crucial. Nat Resour Forum 22(1):37-51
38. Falkenmark M, Molden D (2008) Wake Up to Realities of River Basin Closure. Int J Water Resour Deve 24(2):201-215
39. Fankhauser S (1995) Valuing climate change: the economics of the greenhouse: Earthscan.
40. Fankhauser S (2009) The range of global estimates. In: Parry M, Arnell N, Berry P, Dodman D, Fankhauser S, Hope C, Kovats S, Nicholls R, Satterthwaite D, Tiffin R, Wheeler T (eds) Assessing the costs of adaptation to climate change. International Institute for Environmental and Development and Grantham Institute for Climate Change, London.
41. Farber SC, Costanza R, Wilson MA (2002) Economic and ecological concepts for valuing ecosystem services. Ecol Econ 41(3):375-392

42. Farley J (2012) Ecosystem services: The economics debate. Ecosystem Services 1:40-49
43. Few R (2012) Flood hazards and health: responding to present and future risks. Earthscan: Trowbridge, UK.
44. Fischer EM, Schar C (2010) Consistent geographical patterns of changes in high-impact European heatwaves. Nat Geosci 3(6):398-403
45. Fisher B, Polasky S, Sterner T (2011) Conservation and Human Welfare: Economic Analysis of Ecosystem Services. Environ Resour Econ 48(2):151-159
46. Foley JA, DeFries R, Asner GP, Barford C, Bonan G, Carpenter SR, Chapin FS, Coe MT, Daily GC, Gibbs HK, et al. (2005) Global Consequences of Land Use. Science 309(5734):570-574
47. Fulweiler RW, Rabalais NN, Heiskanen AS (2012) The eutrophication commandments. Mar Pollut Bull 64(10):1997-1999
48. Füssel H-M, Heinke J, Popp A, Gerten D (2012) Climate Change and Water Supply. In: Edenhofer O, Wallacher J, Lotze-Campen H, Reder M, Knopf B, Müller J (eds) Climate Change, Justice and Sustainability, Netherlands: Springer. pp 19-32
49. Galaz V (2012) Geo-engineering, governance, and social–ecological systems: critical issues and joint research needs. Ecology and Society 17:24
50. Gale S, Hoare P (2012) The stratigraphic status of the Anthropocene. The Holocene 22(12):1491-1494
51. Gillings MR (2013) Evolutionary consequences of antibiotic use for the resistome, mobilome and microbial pangenome. Front Microb 4:4
52. Gillings MR, Stokes H (2012) Are humans increasing bacterial evolvability? Trends Ecology Evol 27:346-352
53. Gleick PH, Palaniappan M (2010) Peak water limits to freshwater withdrawal and use. Proc Natl Acad Sci 107(25):11155-11162
54. Gonzalez PL, Polvani LM, Seager R, Correa GJ (2013) Stratospheric ozone depletion: a key driver of recent precipitation trends in South Eastern South America. Climate Dynam. 1-18 doi:10.1007/s00382-013-1777-x.

55. Grandjean P, Landrigan PJ (2006) Developmental neurotoxicity of industrial chemicals. Lancet 368(9553):2167-2178
56. Grandjean P, Bellinger D, Bergman Å, Cordier S, Davey-Smith G, Eskenazi B, Gee D, Gray K, Hanson M, Van Den Hazel P, et al. (2008) The Faroes Statement: Human Health Effects of Developmental Exposure to Chemicals in Our Environment. Basic Clin Pharmacol Toxicol 102(2):73-75
57. Gruber N, Galloway JN (2008) An Earth-system perspective of the global nitrogen cycle. Nature 451(7176):293-296
58. Handoh IC, Lenton TM (2003) Periodic mid-Cretaceous oceanic anoxic events linked by oscillations of the phosphorus and oxygen biogeochemical cycles. Global Biogeochem Cycles 17(4):1092
59. Harnik PG, Lotze HK, Anderson SC, Finkel ZV, Finnegan S, Lindberg DR, Liow LH, Lockwood R, McClain CR, McGuire JL, et al. (2012) Extinctions in ancient and modern seas. Trends Ecology Evol 27(11):608-617
60. Hill M (2013) Addressing Water Governance Challenges in the Anthropocene. Climate Change Water Governance 54:3-16
61. Hoban S, Vernesi C (2012) Challenges in global biodiversity conservation and solutions that cross sociology, politics, economics and ecology. Biol Lett 8:897-899
62. Hodgson DA (2011) First synchronous retreat of ice shelves marks a new phase of polar deglaciation. Proc Natl Acad Sci 108(47):18859-18860
63. Hoegh-Guldberg O, Mumby PJ, Hooten AJ, Steneck RS, Greenfield P, Gomez E, Harvell CD, Sale PF, Edwards AJ, Caldeira K, et al. (2007) Coral Reefs Under Rapid Climate Change and Ocean Acidification. Science 318(5857):1737-1742
64. Hönisch B, Ridgwell A, Schmidt DN, Thomas E, Gibbs SJ, Sluijs A, Zeebe R, Kump L, Martindale RC, Greene SE (2012) The geological record of ocean acidification. Science 335(6072):1058-1063
65. Hooper DU, Adair EC, Cardinale BJ, Byrnes JEK, Hungate BA, Matulich KL, Gonzalez A, Duffy JE, Gamfeldt L, O/'Connor MI (2012) A global synthesis reveals biodiversity loss as a major driver of ecosystem change. Nature 486(7401):105-108
66. Hughes L (2003) Climate change and Australia: Trends, projections and impacts. Austral Ecol 28(4):423-443

67. IPCC (2007) Climate Change 2007: The Physical Science Basis. Geneva, Switzerland: Summary for Policymakers.

68. Jimenez JL, Canagaratna MR, Donahue NM, Prevot ASH, Zhang Q, Kroll JH, DeCarlo PF, Allan JD, Coe H, Ng NL, et al. (2009) Evolution of Organic Aerosols in the Atmosphere. science 326(5959):1525-1529

69. Kaufman S, Qing C, Levenson N, Hanson M (2012) Transport during and after hurricane Sandy. Rudin Center for Transportation, NYU Wagner Graduate School of Public Service.

70. Kinzig AP, Perrings C, Chapin FS, Polasky S, Smith VK, Tilman D, Turner BL (2011) Paying for Ecosystem Services—Promise and Peril. Science 334(6056):603-604

71. Knowlton K, Rotkin-Ellman M, Geballe L, Max W, Solomon GM (2011) Six Climate Change–Related Events In The United States Accounted For About $14 Billion In Lost Lives And Health Costs. Health Aff 30(11):2167-2176

72. Knutson TR, McBride JL, Chan J, Emmanuel K, Holland G, Landsea C, Held I, Kossin JP, Srivastava AK, Sugi M (2010) Tropical cyclones and climate change. Nat Geosci 3:157-163

73. Kosoy N, Brown PG, Bosselmann K, Duraiappah A, Mackey B, Martinez-Alier J, Rogers D, Thomson R (2012) Pillars for a flourishing Earth: planetary boundaries, economic growth delusion and green economy. Curr Opin Environ Sustain 4(1):74-79

74. Lambin EF, Meyfroidt P (2011) Global land use change, economic globalization, and the looming land scarcity. Proc Natl Acad Sci 108(9):3465-3472

75. Lau K, Kim M, Kim K (2006) Asian summer monsoon anomalies induced by aerosol direct forcing: the role of the Tibetan Plateau. Climate Dynam 26(7):855-864

76. Lawson S (2011) Is the future a foreign country? Aust J Politics Hist 57(3):420-433

77. Lewis WM, Wurtsbaugh WA, Paerl HW (2011) Rationale for Control of Anthropogenic Nitrogen and Phosphorus to Reduce Eutrophication of Inland Waters. Environ Sci Technol 45(24):10300-10305

78. Linnenluecke MK, Griffiths A, Winn M (2012) Extreme Weather Events and the Critical Importance of Anticipatory Adaptation and Organizational Resilience in Responding to Impacts. Bus Strategy Environ 21(1):17-32

79. Lobell DB, Burke MB, Tebaldi C, Mastrandrea MD, Falcon WP, Naylor RL (2008) Prioritizing Climate Change Adaptation Needs for Food Security in 2030. Science 319(5863):607-610

80. Lobell DB, Schlenker W, Costa-Roberts J (2011) Climate trends and global crop production since 1980. Science 333:616-620

81. Lurling M, De Senerpont Domis LN (2013) Predictability of plankton communities in an unpredictable world. Freshwater Biology 58(3):455-462

82. Mäder J, Staehelin J, Peter T, Brunner D, Rieder H, Stahel W (2010) Evidence for the effectiveness of the Montreal Protocol to protect the ozone layer. Atmos Chem Physics 10(24):12161-12171

83. Meybeck M (2003) Global analysis of river systems: from Earth system controls to Anthropocene syndromes. Philos Trans R Soc Lond B Biol Sci 358(1440):1935-1955

84. Milly PCD, Wetherald RT, Dunne KA, Delworth TL (2002) Increasing risk of great floods in a changing climate. Nature 415(6871):514-517

85. Molden D (2007) Water for food water for life : a comphrehensive assessment of water management in agriculture. London: Earthscan.

86. Naidoo R, Balmford A, Costanza R, Fisher B, Green RE, Lehner B, Malcolm TR, Ricketts TH (2008) Global mapping of ecosystem services and conservation priorities. Proc Natl Acad Sci 105(28):9495-9500

87. Narita D, Rehdanz K, Tol RS (2012) Economic costs of ocean acidification: a look into the impacts on global shellfish production. Clim Change 113(3–4):1049-1063

88. Nel A (2005) Air Pollution-Related Illness: Effects of Particles. Science 308(5723):804-806

89. Nicholls RJ, Cazenave A (2010) Sea-Level Rise and Its Impact on Coastal Zones. Science 328(5985):1517-1520

90. Nicholls RJ, Tol RSJ (2006) Impacts and responses to sea-level rise: a global analysis of the SRES scenarios over the twenty-first century. Philos Trans R Soc Lond A 364(1841):1073-1095
91. Nilsson M, Persson Å (2012) Can Earth system interactions be governed? Governance functions for linking climate change mitigation with land use, freshwater and biodiversity protection. Ecol Econ 75:61-71
92. Palsson G, Szerszynski B, Sörlin S, Marks J, Avril B, Crumley C, Hackmann H, Holm P, Ingram J, Kirman A, et al. (2013) Reconceptualizing the 'Anthropos' in the Anthropocene: Integrating the social sciences and humanities in global environmental change research. Enviro Sci Policy 28:3-13
93. Pandolfi JM, Connolly SR, Marshall DJ, Cohen AL (2011) Projecting Coral Reef Futures Under Global Warming and Ocean Acidification. Science 333(6041):418-422
94. Parry ML, Rosenzweig C, Iglesias A, Livermore M, Fischer G (2004) Effects of climate change on global food production under SRES emissions and socio-economic scenarios. Glob Environ Chang 14(1):53-67
95. Paytan A, Mackey KRM, Chen Y, Lima ID, Doney SC, Mahowald N, Labiosa R, Post AF (2009) Toxicity of atmospheric aerosols on marine phytoplankton. Proc Natl Acad Sci 106(12):4601-4605
96. Piao S, Ciais P, Huang Y, Shen Z, Peng S, Li J, Zhou L, Liu H, Ma Y, Ding Y, et al. (2010) The impacts of climate change on water resources and agriculture in China. Nature 467(7311):43-51
97. Pielke RA Sr, Marland G, Betts RA, Chase TN, Eastman JL, Niles JO (2003) The influence of land-use change and landscape dynamics on the climate system: relevance to climate-change policy beyond the radiative effect of greenhouse gases. In: Swingland IR (ed) Capturing Carbon and Conserving Biodiversity: The Market Approach, Earthscan Publications. pp 157-172
98. Portmann R, Daniel J, Ravishankara A (2012) Stratospheric ozone depletion due to nitrous oxide: influences of other gases. Philos Trans R Soc Lond B Biol Sci 367(1593):1256-1264
99. Rahmstorf S (2000) The Thermohaline Ocean Circulation: A System with Dangerous Thresholds? Clim Change 46(3):247-256

100. Rahmstorf S (2007) A Semi-Empirical Approach to Projecting Future Sea-Level Rise. Science 315(5810):368-370

101. Rahmstorf S (2010) A new view on sea level rise. Nat Rep Climate Change 4(1004):44-45

102. Reich PB, Tilman D, Isbell F, Mueller K, Hobbie SE, Flynn DF, Eisenhauer N (2012) Impacts of biodiversity loss escalate through time as redundancy fades. Science 336(6081):589-592

103. Repetto R, Easton R (2010) Climate Change and Damage from Extreme Weather Events. Environ 52(2):22-33

104. Rioual P, Andrieu-Ponel V, Rietti-Shati M, Battarbee RW, de Beaulieu J-L, Cheddadi R, Reille M, Svobodova H, Shemesh A (2001) High-resolution record of climate stability in France during the last interglacial period. Nature 413(6853):293-296

105. Rockstrom J, Steffen W, Noone K, Persson Å, Chapin FS III, Lambin E, Lenton TM, Scheffer M, Folke C, Schellnhuber HJ, et al. (2009) Planetary Boundaries: Exploring the Safe Operating Space for Humanity. Ecology Soc 14(2):32

106. Rockstrom J, Steffen W, Noone K, Persson A, Chapin FS, Lambin EF, Lenton TM, Scheffer M, Folke C, Schellnhuber HJ, et al. (2009) A safe operating space for humanity. Nature 461(7263):472-475

107. Ruddiman WF (2003) The Anthropogenic Greenhouse Era Began Thousands of Years Ago. Clim Change 61(3):261-293

108. Ruddiman WF (2013) The Anthropocene. Annu Rev Earth Planet Sci 41(1):45-68

109. Schaeffer M, Hare W, Rahmstorf S, Vermeer M (2012) Long-term sea-level rise implied by 1.5°C and 2°C warming levels. Nature Climate Change 2:867-870

110. Schlegel P, Havenhand JN, Gillings MR, Williamson JE (2012) Individual Variability in Reproductive Success Determines Winners and Losers under Ocean Acidification: A Case Study with Sea Urchins. PloS one 7(12):e53118

111. Schlenker W, Lobell DB (2010) Robust negative impacts of climate change on African agriculture. Environ Res Lett 5(1):014010

112. Schlenker W, Roberts MJ (2009) Nonlinear temperature effects indicate severe damages to U.S. crop yields under climate change. Proc Natl Acad Sci 106(37):15594-15598

113. Seidl R, Brand F, Stauffacher M, Krütli P, Le Q, Spörri A, Meylan G, Moser C, González M, Scholz R (2013) Science with Society in the Anthropocene. AMBIO 42(1):5-12

114. Selin NE, Paltsev S, Wang C, Donkelaar A, Martin RV (2011) Global Aerosol Health Impacts: Quantifying Uncertainties. MIT Joint Program on the Science and Policy of Global Change.

115. Steffen W (2006) The Anthropocene, global change and sleeping giants: where on Earth are we going? Carbon Balance Manage 1(1):3

116. Steffen W, Crutzen PJ, McNeill JR (2007) The Anthropocene: Are Humans Now Overwhelming the Great Forces of Nature. Ambio 36(8):614-621

117. Steffen W, Grinevald J, Crutzen P, McNeill J (2011) The Anthropocene: conceptual and historical perspectives. Philos Trans R Soc Lond A 369(1938):842-867

118. Steffen W, Persson Å, Deutsch L, Zalasiewicz J, Williams M, Richardson K, Crumley C, Crutzen P, Folke C, Gordon L, et al. (2011) The Anthropocene: From Global Change to Planetary Stewardship. Ambio 40(7):739-761

119. Sterling SM, Ducharne A, Polcher J (2013) The impact of global land-cover change on the terrestrial water cycle. Nature Clim Change 3:385-390

120. Stern N (2007) The Economics of Climate Change - the Stern Review: Cambridge University Press. Cambridge: UK.

121. Stoeckl N, Hicks CC, Mills M, Fabricius K, Esparon M, Kroon F, Kaur K, Costanza R (2011) The economic value of ecosystem services in the Great Barrier Reef: our state of knowledge. Ann N Y Acad Sci 1219(1):113-133

122. Tengberg A, Fredholm S, Eliasson I, Knez I, Saltzman K, Wetterberg O (2012) Cultural ecosystem services provided by landscapes: Assessment of heritage values and identity. Ecosystem Services 2:14-26

123. Tickell C (2011) Societal responses to the Anthropocene. Philos Trans R Soc Lond A 369(1938):926-932

124. Tol RS (2002) Estimates of the damage costs of climate change. Part 1: Benchmark estimates. Environ Resour Econ 21(1):47-73

125. Tollefson J (2012) Hurricane sweeps US into climate-adaptation debate. Nature 491(7423):167-168

126. Tsigaridis K, Krol MC, Dentener FJ, Balkanski Y, Lathière J, Metzger SM, Hauglustaine D, Kanakidou M (2006) Change in global aerosol composition since preindustrial times. European Geophysical Society (EGU).

127. van der Leeuw SE (2008) Climate and Society: Lessons from the Past 10 000 Years. Ambio 37(sp14):476-482

128. Velders GJM, Andersen SO, Daniel JS, Fahey DW, McFarland M (2007) The importance of the Montreal Protocol in protecting climate. Proc Natl Acad Sci 104(12):4814-4819

129. Veldkamp A, Polman N, Reinhard S, Slingerland M (2011) From scaling to governance of the land system: bridging ecological and economic perspectives. Ecology Soc 16(1):1

130. Vorosmarty CJ, Green P, Salisbury J, Lammers RB (2000) Global Water Resources: Vulnerability from Climate Change and Population Growth. Science 289(5477):284-288

131. Vorosmarty CJ, McIntyre PB, Gessner MO, Dudgeon D, Prusevich A, Green P, Glidden S, Bunn SE, Sullivan CA, Liermann CR, et al. (2010) Global threats to human water security and river biodiversity. Nature 467(7315):555-561

132. Wetzel FT, Kissling WD, Beissmann H, Penn DJ (2012) Future climate change driven sea-level rise: secondary consequences from human displacement for island biodiversity. Glob Chang Biol 18(9):2707-2719

133. Wilson MA, Carpenter SR (1999) Economic valuation of freshwater ecosystem services in the United States: 1971–1997. Ecol Appl 9(3):772-783

134. Worm B, Branch TA (2012) The future of fish. Trends Ecology Evol 27(11):594-599

135. Young OR, Steffen W (2009) The Earth System: Sustaining Planetary Life-Support Systems. Principles of Ecosystem Stewardship. Chapin FS, Kofinas GP, Folke C (eds) New York: Springer. pp 295-315

136. Zalasiewicz J, Williams M, Smith A, Barry TL, Coe AL, Bown PR, Brenchley P, Cantrill D, Gale A, Gibbard P, et al. (2008) Are we now living in the Anthropocene? GSA Today 18(2):4-8

137. Zalasiewicz J, Williams M, Steffen W, Crutzen P (2010) The New World of the Anthropocene. Environ Sci Technol 44(7):2228-2231
138. Zalasiewicz J, Williams M, Fortey R, Smith A, Barry TL, Coe AL, Bown PR, Rawson PF, Gale A, Gibbard P, et al. (2011) Stratigraphy of the Anthropocene. Philos Trans R Soc Lond A 369(1938):1036-1055
139. Zhang M, Song Y, Cai X (2007) A health-based assessment of particulate air pollution in urban areas of Beijing in 2000–2004. Sci Total Environ 376(1):100-108

Citations

CHAPTER 1

Jure Kokošin and Andrej Gosar, "Seismic Microzonation of Breginjski Kot (NW Slovenia) Based on Detailed Engineering Geological Mapping," The Scientific World Journal, vol. 2013, Article ID 626854, 12 pages, 2013, doi:10.1155/2013/626854.

CHAPTER 2

Anwar Abdullah, Shawki Nassr, and Abdoh Ghaleeb, "Remote Sensing and Geographic Information System for Fault Segments Mapping a Study from Taiz Area, Yemen," Journal of Geological Research, vol. 2013, Article ID 201757, 16 pages, 2013. doi:10.1155/2013/201757.

CHAPTER 3

M García-Talavera, A García-Pérez, C Rey, and L Ramos, Mapping Radon-prone Areas using **γ**-radiation Dose Rate and Geological Information, doi:10.1088/0952-4746/33/3/605.

CHAPTER 4

Said Attia al Hagrey, Daniel Köhn, and Wolfgang Rabbel, Geophysical assessments of renewable gas energy compressed in geologic pore storage reservoirs, doi:10.1186/2193-1801-3-267.

CHAPTER 5

Azusa Nishizawa, Kentaro Kaneda, Yasutaka Katagiri and Mitsuhiro Oikawa, Wide-Angle Refraction Experiments in the Daito Ridges Region at the Northwestern End of the Philippine Sea Plate, doi:10.1186/1880-5981-66-25.

CHAPTER 6

Jose L Palma, Leah Courtland, Sylvain Charbonnier, Riccardo Tortini and Greg A Valentine, Vhub: a Knowledge Management System to Facilitate Online Collaborative Volcano Modeling and Research, doi:10.1186/2191-5040-3-2.

CHAPTER 7

Michael R Gillings and Elizabeth L Hagan-Lawson, The cost of living in the Anthropocene, doi:10.1186/2194-6434-1-2.

Index